may be applied to the grinding of Corn, Flint, Manganesse or other Matter, also to rolling, slitting forging and battering Iron and other [...] other Motion that may be required.

Figure 1st. Is a Retort with its Neck (A) and a Stop Cock (A) I make use of two of these Retorts in order to keep the Engine working with one of them whilst the other is cleansed of the Coaks and Ashes of the Materials used for procuring Inflammable Air.

Figure 2.d Is a Receiver wherein the Air from either of the Retorts is collected and cooled by means of a Circumambrient Cistern of water BB and carried by a small pipe B into the Pump for supplying the Engine with Inflammable Air.

Figures 3.3.d Are the Air pumps which by means of an Admission Valve through which the respective Airs are to be drawn into the Cistern of the Inward Pipe CC (left open at the bottom) and another Valve where the Airs are to pass out of the Cistern into the Compressers does upon the alternate raising and falling of the outward pipes DD (filled with water) convey the Airs into their respective Compressers by being successively raised and fallen by means of two Beams 5.5.

Figures 4.4.4. Are Compressers which being filled in their lower parts with water and the respective Airs forced into them by means of their adjoining Pumps do according to the altitude of the Pipes rising up to their Cistern compress the Air and force it into the Exploders with a power proportioned to such altitude In some cases I chuse to use forcing Pumps and Bellows for performing this Operation.

Figures 5.5. Two Beams wrought by two Esses or portions of Circles fixed on one of the Axis of the Engine Wheel.

Figures 6.6.6. Three Stop Cocks to regulate the Admission of Inflammable Air, Common Air and Water, into the Exploder.

Figure 7. The Exploder with a Pipe (E) which introduces Inflammable Air, a Pipe (F) which introduces Common Air and a Pipe (G) which introduces Water into this Vessel, at whose Mouth or outlet (H) (upon the approach of Flame) the combined Streams of Airs and Water do issue out with amazing force and velocity against.

Figure 8. The Fly Wheel on whose Axis a Pinion being placed turns a toothed Wheel of proper dimensions and on the same Axis are affixed.

Figures 9.9. Two Metallic Esses or Portions of Circles set at right Angles from each other so as by rotation of the Axis whereon they be fastened they bear upon the Wheel (LL) in the Plug frame (KK) suspended to the Beams 5.5 and thereby work those Beams two strokes each in every rotation.

Figure 10. Is the second Wheel in the Engine wrought by the Pinion upon the Fly Wheel and upon whose Axis is fixed another Pinion which working upon another toothed Wheel of any proper diameter may upon the Axis thereof carry either Cogs or Esses for lifting Hammers, Barrels for Ropes to wrap upon, Cog Wheels for communicating Motion to Grinding or Spinning Machinery or rolling slitting and pressing Metals or other Substances and also for any kind of Motion whatsoever.

[...]our sort of Engines and Machines worked upon these [...] sixpence wrought with very little Fuel, and managed with little [...] Machinery hitherto invented.

Drawn on Stone by Malby & Sons

[...]d William Spottiswoode.
[...] Majesty. 1855.

First published in England 1982 by
Eton Publishing,
50 High Street,
Eton, Berks.

ISBN 0 9508076 0 5.

Designed and photoset by —
Arrowhead Publishing,
Unit 3, Park Works,
Kingsley, Hampshire.

Printed by Broglia Press   Denmark Road, Winton, Bournemouth, Dorset.

POWER FOR THE FLEET

The History of British Marine Gas Turbines

By

C.E. Preston

ACKNOWLEDGEMENTS

A great part of the material in this book is drawn from the large number of contemporary papers contributed by those who were engaged in the work, to the Institute of Marine Engineers, the American Society of Mechanical Engineers and the Journal of Naval Engineering.

We owe a particular debt to Vice-Admiral Sir Allan Trewby, who was closely associated with the early development, and guided its progress over a number of years, and to his successors at Bath, not least, of course, to Vice-Admiral Sir George Raper, who was finally responsible for leading the Navy into the gas turbine age in 1967.

We are equally indebted to those who carried out, and chronicled, the work of development in Metropolitan-Vickers, Bristol Aero-Engines, Bristol-Siddeley, and Rolls-Royce. There is no space to list all those responsible, but mention must be made of Dr D.M. Smith and Mr. F.R. Harris, of Metropolitan-Vickers; of the late Mr. A.H. Fletcher, who designed the RM 60 and later introduced the Tyne; of Mr. B.G. Markham, who launched the Proteus on its long career; and, covering the whole period from 1960 onwards, of Mr. W.H. Lindsey, and his successor, Mr. B.H. Slatter, first in Bristol-Siddeley and later in Rolls-Royce, who, between them, led marine gas turbine development for twenty years.

On a personal note, I am most grateful to the many colleagues in Rolls-Royce, as well as in the Ministry of Defence, BHC, and Vosper, who have helped unstintingly in providing data for the book. My special thanks are due to Cdr. Craig, the editor of the *Journal of Naval Engineering,* who provided a wealth of information; to Mr. Malcolm Bromwich, who furnished so many of the illustrations; and to Mr. W.J.R. Thomas, Mr. B.H. Slatter, Mr. B.G. Markham, and Rear-Admiral O'Hara who took the trouble to read the manuscript and to give me much excellent advice, on historical matters, on the niceties of engineering, and on the spelling of the English language.

for Joy

CONTENTS

Front Cover — Sectional drawing of Rolls-Royce Industrial Olympus Gas Generator, part of a Marine Proteus, and HMS *Invincible*.

Back Cover — Rolls-Royce Tyne Gas Turbine.

End papers — The end papers show John Barber's patent of 1791, the first known patent for a gas turbine. Hardly recognizable as such to modern eyes, the machine still has all the basic components of today's gas turbines.

FOREWORD

by Vice-Admiral Sir George Raper KCB

It gave me enormous pleasure to be asked to write a Foreword to such an ably and knowledgeably written story of gas turbines in the Royal Navy. Their adoption in place of steam plant was the consummation of a development policy started in 1942 and followed consistently by successive Engineers-in-Chief and Directors of Marine Engineering in the Admiralty until 1967, when it came to fruition in the new classes of frigate and destroyer and in H.M.S. *Invincible.*

Those who believed in talking about thermo-dynamics foretold that the efficiency of gas turbines in the Carnot Cycle was bound eventually to be better than the Rankine Cycle of the Steam Plant. But to those of us who were brought up on steam the final decision was inevitably difficult. Steam plant could be kept going by hard work, in rather unpleasant conditions of heat and humidity, for long periods, provided the minor faults and leaks were attended to early enough. This may have built character in the operators and maintainers, but it also left an abiding memory of nights in harbour spent in hard work by Artificers and Stokers or Mechanics who still had almost endless chores of cleaning up to perform.

Gas turbines gave the promise of a much better quality of work and less of it on board. The aircraft-derived gas turbine had been designed with quick replacement and factory rectification of faults in mind. The work of servicing *in situ* was also much reduced, provided they ran long enough without trouble.

After the Defence Review of 1965, which cut off the future of aircraft carriers in the Fleet, new classes of frigate and destroyer were called for and the need for greater 'availability' and fewer men in the ships was emphasised. Some measure of both was offered by a change to gas turbines. Moreover the development of aircraft jet engines had reached the stage when their fuel-economy was a good as, or perhaps even better than, the economy of current naval steam plant. The RN was also faced with the need to build three large cruisers to carry helicopters — and possibly, it was rumoured, the Harrier.

Naval machinery needs high power for full speed, seldom used, and good fuel economy at a low proportion of full power for cruising conditions.

These requirements are peculiar to warships. Merchant ships were being fitted with Diesel engines increasingly so the market for steam plant derived for naval purposes was a small one. It was evident that the high-grade talent needed in industry, for the development of plant of reasonably reliable characteristics, could only be justified by the attraction of production orders to the shops, so the size of the market was very important.

It so happened that the power needed on each shaft in the new ships could be provided by gas turbines, derived from aircraft engines and developed for electric generation in power stations. There then had to be found an engine of good fuel-economy for cruising power. This book tells the story.

The setting up of the Rolls-Royce Industrial and Marine Division gave the assurance that the necessary talent would be available to solve the problems of the new ships.

It therefore only remains to express the gratitude of the present generation of Engineer Officers in the RN to the senior officers of the quarter-century before 1967, to Rolls-Royce, and in particular to those in the Marine Division, and to the people in Metropolitan Vickers (later AEI and finally GEC) who carried the burden of development, particularly through the combined steam and gas turbine plant of the Guided Missile Destroyers of the 'County' Class and thus made the all-gas turbine propulsion of modern warships, both possible and popular.

INTRODUCTION

The start of a new era. **Turbinia** at the Naval Review, 1897.

For over 100 years, the navies of the world relied on steam machinery for propulsion, and it need hardly be stressed that its introduction completely revolutionised naval warfare and peacetime operations. Nevertheless, the naval authorities were extremely reluctant to accept steam when it was first introduced, and they continued to view it with the greatest suspicion for 25 years, before the decision was finally taken, in the early 1870s, that the Royal Navy should abandon sail and rely henceforth entirely on steam.

From the time of its general introduction in the mid-19th Century, steam plant was steadily improved and refined, efficiency was raised, and new systems were developed to meet the varying needs of the Fleet. It is a far cry from the old box boiler of the 1840s, operating on seawater at a pressure of 10 pounds per square inch (psi), with a push-and-pull engine running at 30 revolutions per minute (rpm) to the 400 psi steam turbine of Second World War vintage. In the 1940s further advances in machinery design and in steam pressures and temperatures were already practicable. Steam did serve the navies of the world very well, providing a universal, efficient, and highly reliable means of propulsion for virtually every type of ship during the great industrial expansion of the 19th and early 20th centuries, and throughout two world wars.

It is therefore all the more surprising that in the space of some 30 years, between 1945 and 1975, steam plant, which was unchallenged at the beginning of that period, should have been completely superseded by gas turbine machinery in all new designs of virtually every major navy.

Of course, a large proportion of the world's warships at sea today are still steam-powered, since the life of an average naval vessel is, even in theory, 25-30 years, and in practice usually a good deal longer. But it now seems highly unlikely that any navy will again embark on a new ship-design incorporating steam machinery, apart from relatively few nuclear submaries and a much smaller number of nuclear-powered cruisers and aircraft carriers.

Since the first introduction of steam, there had been one major change in ship machinery prior to the advent of the gas turbine. This was the introduction of the steam turbine at the turn of the century, and it is interesting to draw a parallel between the two developments.

Perhaps the thing that most strikes us about the steam turbine is the speed with which it was adopted, by a hitherto sceptical, and usually highly conservative world. Before Sir Charles Parsons

S.S. **Turbinia**. *"...within a few years, the steam turbine had completely displaced the reciprocating engine..."*

first introduced his new engine in the yacht *Turbinia,* and brought it a blaze of publicity by demonstrating her unmatchable performance in the lines of the Naval Review in 1897, naval authorities had shown little interest in the new system. Yet by 1905 we find the Admiralty Committee on Designs, presided over by Admiral Fisher, virtually deciding that all future ships of the Royal Navy should be fitted with steam turbines. We may find a clue to this decision in the words of the then First Lord, at the time of the laying down of H.M.S. *Dreadnought:*

'While recognising that the steam-turbine system of propulsion has, at present, some disadvantages, yet it was determined to adopt it because of the saving in weight and reduction in number of working parts and reduced liability to breakdown, its smooth working, ease of manipulation, saving in coal consumption at high powers, and hence boiler-room space, and saving in engine-room complement, and also because of the increased protection provided for with this system due to engines being lower in the ship — advantages which more than counterbalance the disadvantages. There was no difficulty in arriving at a decision to adopt turbine propulsion from the point of view of sea-going speed only. The point which chiefly occupied the Committee was the question of providing sufficient stopping and turning power for purposes of quick and easy manoeuvring.'

Sixty years later, almost exactly the same arguments were being put forward in justification of the much more far-reaching revolution brought about by the introduction of the gas turbine, and the same problem of manoeuvring — or at least of how best to provide astern power — was exercising people's minds.

By contrast with the steam turbine, however, the gas turbine revolution was much slower in coming to its full fruition. The first experimental gas turbine went to sea in the Royal Navy in 1947; the world's first warship to rely entirely on gas turbine propulsion, H.M.S. *Grey Goose,* commissioned in 1953; but it was not until 1967 that the British, first among the world's major navies, made the decision to rely on gas turbines as the sole propulsion engines for future warships.

It was a decision that we can see, with hindsight, could have been made considerably earlier, since the engines eventually chosen were already available in 1960. But at this time the only gas turbines being considered for propulsion of larger vessels were relatively unsophisticated and limited in power, having been designed as a supplement to conventional steam machinery; and the final adoption of all-gas turbine plant meant, in the end, accepting not only a new type of prime mover, but a whole new machinery philosophy borrowed from the quite alien world of the aircraft.

Even in the late 1950s and early 1960s, when some gas turbine machinery was already in service in the Royal Navy, most marine engineers and ship designers, both in Britain and elsewhere, were quite adamant that any gas turbines for use in major warships must be robust and relatively heavy 'marine' engines, purpose-built for warship use, and that it would be quite impracticable to use the lightweight aircraft type of engine, which was then making its debut in motor torpedo boats and other small craft. However, it was soon to be realized that, if the Royal Navy wanted an engine of high power and efficiency, and also reasonably compact, there were enormous advantages to be gained, both in development cost saving and in reliability, if marine engineers could build on the much greater amount of work which had already been done in the field of aero-engines.

So the Navy's decision, when it came, was based on aero-derived engines. Other navies were moving along the same lines, or were soon to follow the British lead, and today the lightweight aero-derived gas turbine is used universally in all types of warship from small high-speed craft to large aircraft carriers.

Although other countries have played a major part in the development of both marine and land-based gas turbines, of many types and sizes, it is still true to say that the marine gas turbine did originate in Britain, that it was pioneered by the Royal Navy, and that it has continued, in its world-wide application, to be in large part a product of British design and development.

Use of gas turbines is, of course, not necessarily confined to naval craft, and in the early days of their development many people envisaged their being used widely for propulsion of merchant ships. This may still come about, but a number of factors have combined to make the present generation of gas turbines uneconomic for most merchant ship applications, and current installations are confined to a few specialized cases. While this book must therefore make some reference to merchant ship applications, and the story could hardly be told without the mention of the work done in other countries, our main concern is with British naval gas turbines.

Amongst these, the majority today are used for ship-propulsion. A considerable number of smaller engines were developed in the earlier days for auxiliary use in electric generator sets, and units such as the Allen 500 kw and 1000 kw, the Ruston TF, and the small Rover and Centrax gas turbines have all been employed in earlier ships, while the Rover has also been used to power portable fire pumps. A few Allen gas turbines are still in service in the 'Tribal' class frigates, and a number of Rover engines are installed in fast patrol boats. The Rover, indeed, was used for a time as a propulsion engine, first in 1950, for demonstration in an ex-Torpedo Recovery Vessel, *Torquil,* and thereafter, for some years, in regular service in a harbour launch attached to the Trials Squadron of Coastal Forces. However, we still seem to be some way from evolving an economic propulsion gas turbine for such small craft, and if it ever does appear, it will almost certainly be merely as a very minor application of an established road vehicle engine.

As far as auxiliary power is concerned, the high-speed diesel has now largely replaced the gas turbine in current frigates and destroyers, and we have yet to develop a really satisfactory gas turbine for this job, which would confer the benefits of low noise and easy maintenance, without incurring too great a penalty, either in high specific cost or in poor part-load efficiency.

During the last 20 years, interest has been focussed almost entirely on propulsion gas turbines for sea-going craft, and it is of these engines that we have endeavoured in this book to trace the development and application throughout their short history, from the day in 1947 when the first prototype engine went to sea in a converted motor gunboat, up to the present time, when they power the ships of 25 of the world's navies, and are installed in a range of vessels extending from 7-ton hovercraft up to the 20,000 ton *Invincible* class aircraft carriers.

ORIGINS

If we ignore the invention of Hero of Alexandria in 60 AD — sometimes claimed as the first gas turbine, but really a steam turbine — we can date the concept of the gas turbine from the year 1791, when an Englishman, John Barber, took out British Patent No. 1833 for "A Method of Rising Inflammable Air for the Purpose of Producing Motion, and Facilitating Metallurgical Operations".

Although it is highly unlikely that Barber ever built, still less operated his machine, he must be given the credit for having included all the basic elements of a gas turbine in his proposal — an achievement all the more remarkable when it is realized that over 100 years were to elapse before a practical machine was to be constructed which was capable of actually running, however inefficiently. Barber's idea was to use coal gas as a fuel, and his proposal included reciprocating compressors to pump gas and air into a combustion chamber. Here the gas was ignited, and the hot exhaust used to drive a turbine wheel.

Barber's was the first of a long line of attempts to do two things: first, to convert the energy in a hot fluid — gas or steam — into shaft power without having to use a complicated array of pistons, piston rods, and connecting rods, all being accelerated and decelerated rapidly in opposite directions, to drive a crankshaft. If a turbine wheel could replace the piston-cylinder-crankshaft arrangement, it would also have the advantage of allowing continuous application of power, rather than the intermittent thrust on each stroke of the steam piston, or the still more violent force of successive near-explosions in the internal combustion piston engine.

The second goal of the gas turbine inventors was to find a way of using the hot furnace gases themselves as the working fluid in the engine, rather than going through the highly complicated process of transferring their energy to a separate fluid, in the form of steam, and then recovering this for use again in the engine cycle — a process involving the use of boilers and condensers and feed tanks, and a host of auxiliary pumps.

The internal combustion piston engine did avoid the need for steam; but only the gas turbine could solve both problems. Unfortunately, it had two problems of its own, which were to delay its practical realization until the 1930s. One stemmed from its very advantage of continuous combustion, since this produced sustained high temperatures, involving the use of heat-resisting materials which were not available in the early days. Even with such materials, it was necessary to cool the hot gases by some means before they were introduced into the turbine. Barber proposed to do this by injecting water into the gas stream. This was a common arrangement in many of the innumerable gas turbine designs of the 19th century, but the introduction of additional air for cooling, which is, of course, universal today, was first proposed by a Frenchman, M. Breeson, in 1837.

The other problem which faced the early inventors was their lack of knowledge of aerodynamics, and the consequent lamentable inefficiency of their turbines and compressors. Barber, and some of the other early inventors, avoided the problem to some extent by using reciprocating compressors, rather than the rotating axial or centrifugal types which we know today; but throughout the 19th century all the proposed designs suffered from this drawback to such an extent that it was not until 1900 that a gas turbine was actually built and run, by the German, Dr Stolze, and even this unit appears to have been so inefficient that it produced no useful output, using all its available turbine power to drive its own compressor.

It may be appropriate here to say a word about basic gas turbine principles, for the benefit of those unfamiliar with these machines. As in all internal combustion engines, four pro-

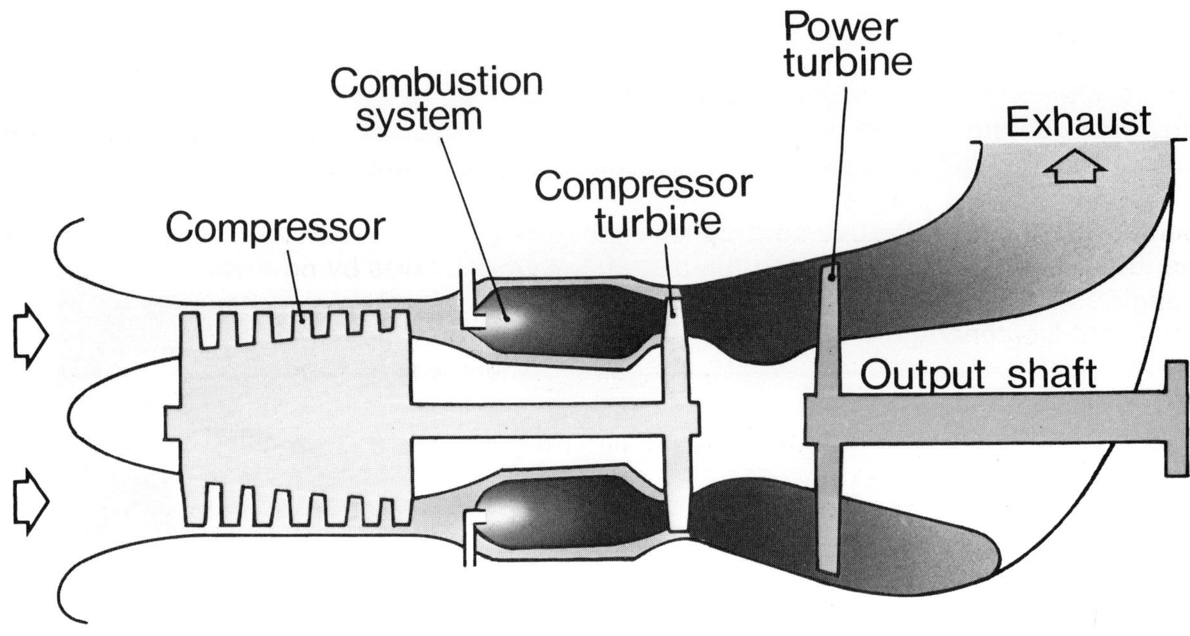

The basic components of the gas turbine. Although later engines incorporate many variations and refinements, the cycle used is essentially the same.

cesses are involved: first, air is drawn in and compressed; second, it is mixed with fuel and combustion takes place; third, the resultant hot gas is partially expanded to provide the power for the compression process; fourth, the remaining energy in the hot gas provides output power. However, whereas the piston engine performs all four functions consecutively in the same space — using its cylinder and piston, first as an air suction, then as an air compressor, then as a combustion chamber, then as a driver — in the gas turbine each function is carried out in a separate section of the engine, so that all four can run concurrently. It therefore contains, in its basic form, a compressor to draw in and compress the air; a combustion chamber, into which the fuel is pumped, mixed with the air, and burnt; and a turbine, driven by the hot gases from the combustion chamber, which provides the power, both for the compressor and for the output shaft.

In practice, two turbines are used in marine gas turbines (as well as in many industrial units). This is because, although the power required to drive a ship's propeller does fall rapidly as the speed of the propeller is reduced, the power available from the gas turbine falls even more rapidly as the speed of the compres-

sor is reduced; so if the same turbine were used to drive both the compressor and the ship's propeller, and the engine were designed to produce the right amount of power at high speed, then at lower speeds the power produced by the engine would be insufficient to drive the propeller. It is therefore necessary to split the turbine portion of the engine, using one section to drive the compressor, on a common shaft, and another quite independent section to drive the output shaft, with each section running at the appropriate speed for the power requirement of the moment.

So much for the theory. It will be appreciated that, in practice, if the efficiency of the compressor and turbine is reduced to a certain level, all the energy fed into the engine by burning fuel will be used up in driving the compressor. If so, the whole device may be decorative and highly ingenious, but it will serve no useful purpose. If the efficiency falls below this level, it will not run at all; if the efficiency rises, there is surplus power available, and we start to have a useful prime mover.

This was the point reached in the early 1900s, but although the principle of the gas turbine had been successfully demonstrated, progress was still very slow, and practical applications were confined to a few industrial machines for electrical generation. In 1939, Brown Boveri, who had worked on gas turbine development for many years, commissioned a 4000 Kw gas turbine generating set at Neuchatel which reached an efficiency of 18 per cent, and this represented the peak of achievement up to that time. But the outbreak of war, and the aircraft jet-engine work already initiated by Sir Frank (then Flight-Lieutenant) Whittle in 1930 were to have a dramatic effect on the whole story of gas turbine development.

It is a commonplace today that virtually all naval gas turbines are of the aero-derived type, either developed from an existing aero-engine, or at least using largely aero-engine components and based on aero design philosophy. This was not the case in the early days, as far as major warships were concerned. The first gas turbines to go to sea in larger ships such as frigates and destroyers were those installed in the British Tribal class, the German *Köln* class, and the Russian *Petya* class, all of which commissioned about the same time, in 1960-61. With the possible exception of the British units, these engines owed little to aero-engine development, and their design was based more on steam turbine philosophy. When aero-derived engines were first introduced for this type of ship in the late 1960s it was in part through the efforts and enterprise of the gas turbine man-ufacturers and the shipyards, and they made their debut in smaller navies, in the teeth of considerable opposition from established marine engineering opinion. And although the Royal Navy can rightly be said to have pioneered the use of gas turbines, and today uses them exclusively for all new major warships, the British Admiralty was by no means the first to accept today's philosophy of adapting existing aero-engine designs.

There was one field, however, in small coastal craft, where the British had in fact a well-established tradition of using converted aero-engines, which dated from pre-war years. Rolls-

The Rolls-Royce Merlin not only powered the **Hurricane** and the **Spitfire**, but was used in many MTBs during the War.

BELOW: **Celerity**, built originally as an Australian Air-Sea Rescue boat, used four **Hercules** engines, and achieved 36 knots on trials.

Royce and Napiers supplied a considerable number of Merlin, Griffon and Lion engines for a wide range of motor torpedo boats — over 100 marinised Merlins were installed in Coastal Forces craft between 1938 and 1941.

The Packard engine, which came later from the United States in great numbers, and which was to remain the standard petrol engine for all our MTBs until the early 1960s, was also of aero-engine type.

At a later date, even radial air-cooled engines were used at sea. At first sight it may seem incongruous that, with all the water in the world available, any boat-designer should choose an air-cooled rather than a water-cooled engine; but the radial engine did have some advantage in light weight and compactness. Its short length made installation easier in some machinery arrangements, and the large frontal area, which could be a disadvantage in aircraft, presented no particular problem for the boat designer. The water-cooled engine was not, of course, quite the obvious choice which it might at first appear to be, since it could not use salt water directly for cooling, and had to incorporate a heat exchanger, but in the long run it was to be accepted as the standard type of unit for all small craft.

Nevertheless, when the Admiralty decided during the war that, to provide an alternative to the Packard engine, with its ever-vulnerable supply line from the United States, they should develop a British unit for Motor Torpedo Boats (MTBs), one choice was the Bristol Hercules, of 1675bhp, and four of these were installed in a Fairmile 'D' motor torpedo boat for trials. Later in the war, Bristol again used four Hercules in a long-range air-sea rescue boat named *Celerity,* which was ordered by the Australian Government. Although, with the ending of the war, the project was cancelled, *Celerity* was completed and carried out successful sea trials with the RN before she was scrapped in 1946.

In the immediate post-war years the rapid rundown of British coastal forces left the RN with large stocks of the well-proven Packard engine, and they also had access to the equally successful German Mercedes-Benz diesel. There was little cause to continue experimenting with aircraft piston engines, and by that time more exciting prospects were appearing in the form of the gas turbine.

Experience in coastal forces had demonstrated that aircraft piston engines could operate satisfactorily in the MTBs with relatively little modification, and during the war years gas turbine development had been entirely in the hands of the Air Ministry and the aero-engine industry. It was natural, therefore, that, whatever long-term plans might be developed, the Admiralty should embark initially on trials of an aero-type gas turbine in an MTB, as a first experimental step along the new road.

It is indeed remarkable that the Navy should have decided to turn its attention to the new type of engine so early, and at a period of very great tension in the war at sea. It was in July 1942 that the Engineer-in-Chief started discussions with Metropolitan-Vickers on the possibility of developing a marine propulsion unit based on an aircraft-type gas generator, or jet engine. In August 1943 a contract was placed with Metropolitan-Vickers for the production of three engines, and four years later, on 14th July, 1947, almost exactly 50 years after the historic appearance of the steam turbine in the *Turbinia,* the world's first marine gas turbine commissioned for trials in Motor Gunboat *No. 2009.*

FIRST ATTEMPTS

The forerunner of a prolific line of marine gas turbines was known as the Gatric. It was based on the F.2 jet engine, which had been developed during the war years by Metropolitan-Vickers, coupled to a newly-designed power turbine to convert the energy in the jet thrust to shaft power.

The British thus started with a basic theory — of mating an aero jet engine with a specifically marine power turbine — which has been continued in all the later developments of naval gas turbines, although the practical application has changed almost out of recognition since those early days. In some respects the F.2 was not an ideal engine to choose for a first essay in marine gas turbine development, as it had, even by the standards of those days, a relatively low compression-ratio of 3.5 to 1, resulting in correspondingly low efficiency, and its aero background was limited. But it did have the decisive advantages of being available without further major development expenditure, and of being about the right size for use in an existing Coastal Forces craft. With its four-stage power turbine, the complete unit developed 2500 hp, with an efficiency of 12.8 per cent, equivalent to a fuel consumption of 1.07 lbs/hp.hr.

It had originally been intended to instal two engines in a Fairmile 'D' class MTB, but this involved considerable re-design of the machinery and, in addition, it became apparent towards the end of 1945 that teething troubles which had developed in the Gatric during test-bed running were starting to make the supply of two engines for installation in a ship increasingly doubtful. The Fairmile 'D', being a four-engined boat, was ill-suited to a single trial-engine installation, and it was accordingly decided in early 1946 to use instead one of the three-shaft Camper and Nicholson Motor Gunboats (MGBs). These craft, 116ft long and displacing 100 tons, were originally powered by three Packard petrol en-

gines, and there was sufficient space in the engine room to accommodate the Gatric in place of the centre engine without moving the existing wing engines.

Admittedly this meant mating an experimental gas turbine, whose external casings ran at pretty high temperatures, with two petrol engines in a common engine room, but it did provide a quick and cheap means of getting some initial sea experience with a gas turbine.

It has been argued that, both with the Gatric and with the later and much more ambitious Rolls-Royce RM 60 gas turbine, sea experience, in the somewhat sheltered environment of an experimental trials programme, did little to improve the engineers' knowledge or to advance the cause of gas turbine propulsion. Certainly from the point of view of the development engineer, test-bed running of a new unit is much more rewarding than operation at sea, mainly because, on the test-bed, the development engineer has the whole programme firmly within his grasp, uninterrupted by such considerations as bad weather, and can indulge his thirst for knowledge by garlanding his engine with festoons of instrumentation which just could not be accommodated in a small ship installation. He does have, too, the very genuine advantage of being able to simulate real-life conditions — provided those conditions have already been sufficiently studied at sea to be reproducible on shore — whereas it is notoriously difficult to find 'real-life' conditions in real life just when you happen to want them.

Later, in 1969, when HMS *Exmouth* headed for northern waters to carry out Arctic trials with the first Rolls-Royce Olympus engine, she encountered the most appalling weather, with gales up to force 12 — but was quite unable to find the extreme cold conditions which she was really seeking. Before that, those onboard the Bold class fast patrol boats can recall frustrating weeks and months searching round Britain's storm-tossed coasts for some really short, steep seas, suitable for rough weather trials — to be greeted all too often by the disappearance of the promised wind and the return of a benign and quite useless calm.

Nevertheless, there is still much virtue in testing the real thing — not as a substitute for test-bed running, but as a necessary complement — partly because no test-bed will ever really reproduce sea-going conditions, and partly, perhaps, for psychological reasons.

It is perhaps worth remarking that the United States, with its much greater technical and financial resources, started a programme investigating and developing marine gas turbines in 1940. This was continued through the war years, with the design and construction of three experimental gas turbines for shore trials and two free-piston gas generator projects. But it was not until 1966 that the US Navy first put gas turbines to sea, apart from small auxiliaries, the 200hp engines in the diminutive minesweeping launches and those in the experimental hydrofoil craft, USS *High Point,* which was powered by two British Proteus engines. It was not until 1975, with the appearance of the *Spruance* class destroyers, that they first used gas turbines in major warships.

A wise senior officer of the United States Navy once remarked that the US Bureau of Ships was hampered in its development efforts by having too much money — they could afford to mount large-scale programmes, and explore every possible approach to a problem in great depth, whereas smaller navies, with more limited resources, were forced to make earlier decisions, and concentrate their efforts in limited practical channels. The 'limited resources' approach certainly saves time, and more often than not it seems to produce a sound result. We may argue endlessly about pragmatism in politics, but there is surely a very good case for pragmatism in engineering.

In 1946, nearly a year before the Gatric started sea trials, the British Admiralty, as we shall see, already had two other experimental gas turbine projects under way. It might well have been decided to continue with the broadly-based shore development programme for some years longer before taking the decision to embark on sea trials. But most people would agree today that this was a right decision, however imperfect the engine and the vehicle, and even though the financial restraints of that time meant a rather uneasy partnership between the diesel-burning Gatric and the two petrol engines. Apart from careful attention to engine-

First of a long line. The **Gatric** *gas turbine, based on the Metropolitan-Vickers F.2 jet engine, with the addition of a four-stage power turbine.*

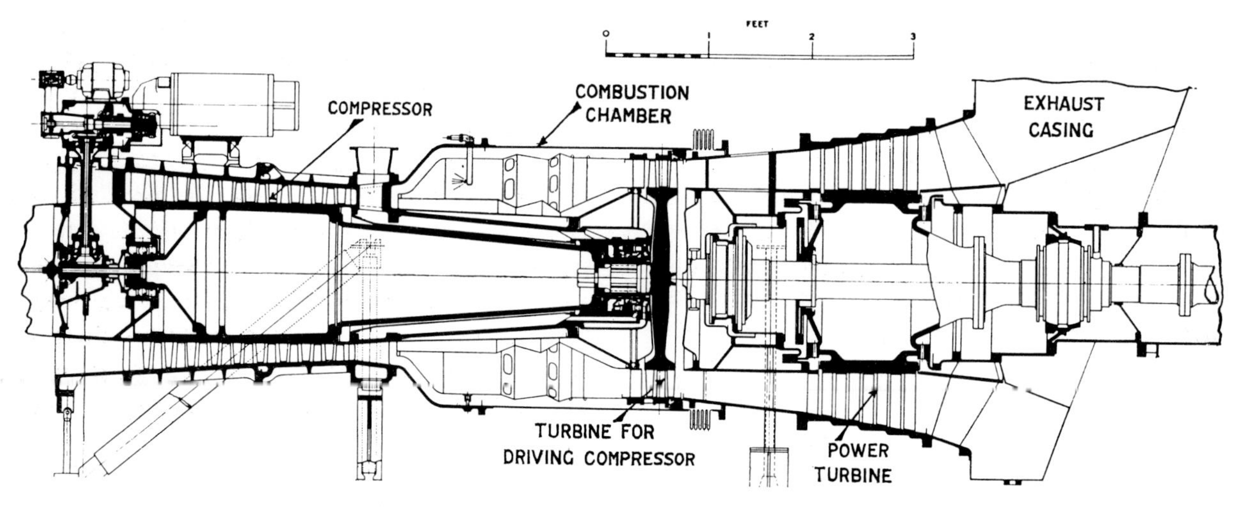

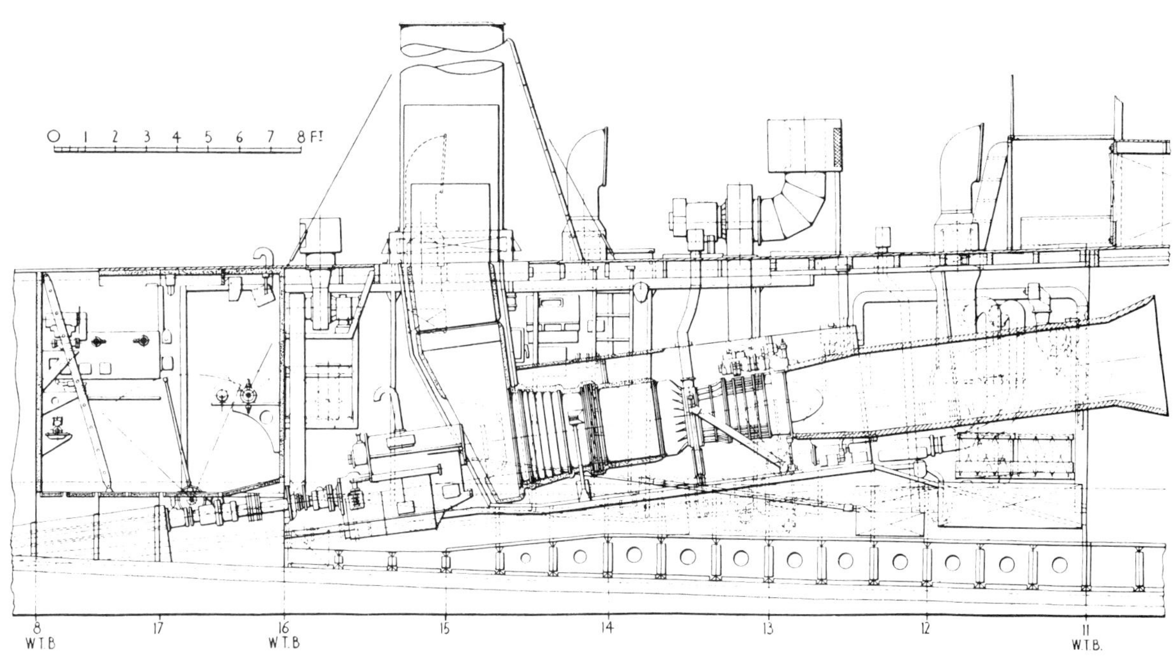

*The **Gatric** in M.G.B. 2009. Intake air was ducted from a sheltered position abaft the bridge, but little attempt was made to exclude salt spray.*

room ventilation, and the fitting of additional fire-extinguishing equipment, the main precaution against fire was the enclosure of the complete gas turbine and exhaust duct in a separately ventilated outer casing — a forerunner of what was to become standard practice, as far as the engine itself was concerned, in nearly all the later units.

In one respect, the installation of the Gatric was certainly not typical of later arrangements. It was not fully appreciated in those early years — not even during the 1950s — that the gas turbine, although able in theory to swallow quite happily considerable quantities of water, would in practice be highly sensitive to salt, and thus allergic to sea water. Salt corrosion, and the build-up of salt deposits on compressor blading, have caused trouble since the earliest days, and it was not until the early 1970s, with the introduction of today's highly efficient multistage filtration systems, that salt-ingestion, if not entirely eliminated, was at least no longer a practical problem.

In *MGB 2009,* no attempt was made to filter the incoming air, but the intake was taken from abaft the bridge. The air was led thence for-

wards and downwards to a settling chamber, where it was hoped, rather over-optimistically, that the seawater droplets entrained in the air would fall to the bottom, in a relatively low-velocity air stream. The gas turbine took its air supply from a point halfway up the rear face of the settling chamber, through a cylindrical duct made of metal-banded balsa wood, 12ft long and weighing over 300 lbs. The length of duct was, of course, dictated by the distance between the back of the bridge and the front of the engine; balsa wood was used for its sound-absorbing properties, and this was in fact the only means employed to reduce noise. By all accounts it was successful in making the engine-room reasonably habitable — in fact, the Gatric engine-room was quieter than that of a standard Camper and Nicholson MGB with three Packard engines — but there was little reduction in the external noise-level, which was described as 'a mixture of a scream, a roar, and a whine, with the scream predominating', and 'most painful... in a position between the funnel and the intake'.

It is perhaps unfortunate that the jet engine's reputation for noiseness should have followed it to sea in this first installation, for there is no doubt that the next generation of sailors was nourished on the myth that gas turbines make an ear-splitting row, whereas, of course, with relatively simple silencing arrangements, the marine installation is extremely quiet. Indeed, in a noise survey carried out later in a modern destroyer, with two Olympus engines running at high power, the highest noise-level measured on the upper deck was in the vicinity of the ventilating fan for the helicopter hangar!

To say that the Gatric installation had its shortcomings is not to condemn it, but rather to demonstrate its value in focussing attention on some of the areas which needed further study. In all, it was an imaginative and highly success-ful first essay with an exciting new type of engine; it gave confidence in the potential for gas turbines in the future, and encouragement to further development. Some of us thought later, when we encountered fresh problems, that the Gatric had been too tenderly 'nursed' during its sea trials, and that more benefit might have been gained for the next stage in the development process, if further extended running under more exacting conditions could have been carried out — including, for instance, rough weather trials.

Perhaps this was so, but it was not a bad launch for a major machinery revolution, and we can claim today that it did start a programme which led to the ultimate development of a comprehensive and highly successful range of marine gas turbines, and the transformation of main machinery in the fleet, both for the Royal Navy and for the other major navies of the world.

One early result of the Gatric trials was to give support to the view, which had many opponents, both then and later, that the aero type of gas turbine did have a part to play in marine applications. Gatric had been chosen for initial trials partly because there was no possibility of producing a purpose-built marine engine without incurring heavy expenditure, and accepting considerable delay in putting a gas turbine to sea. By the time *MGB 2009* started trials in 1947,

H.M.S. **Bold Pioneer.** *Powered by two G.2. gas turbines and two MB518 diesels of 2500 hp, with a conventional hard-chine hull, she achieved 42 knots on trials and proved a good sea boat.*

H.M.S. **Bold Pathfinder,** *the round-bilged sister ship of* **Bold Pioneer.** *The main objectives of the two prototypes "Bold" boats were hull form evaluation and hull stress measurement for a new type of convertible MTB/MGB. The gas turbines were very much taken for granted.*

two other gas turbine development contracts had already been let, one to English Electric for the 6500 hp EL 60A, and the second to Rolls-Royce for the very advanced and, by the standards of those days, very efficient RM 60, of 5400 hp. Both these engines were to be designed *ab initio* as marine units, the first intended for use as a long-life engine in major warships, and the second planned initially as a high performance main propulsion-unit for light craft. Although the RM 60, coming from an aero-engine company which already had considerable experience with jet engines, was based on aero design philosophy, it was in no way an adapted aero engine, while the EL 60 A was clearly based on steam practice, in the belief that, for long-life operation in a big ship, it was essential to use a heavy machine, with plain bearings, able to resist high shock-loads, and intended for permanent installation in the ship.

Before either of these two projects was far advanced, however, a requirement had arisen for gas turbines to be used as boost engines in a new class of convertible Motor Torpedo Boat/ Motor Gunboat, which was being planned as part of the replacement programme for the wartime MTBs and MGBs that were then still in service. The first two boats of the class, *Bold Pioneer* and *Bold Pathfinder,* were planned primarily as development craft, and in particular for hull form evaluation. This was at a time when a fierce but rather fruitless argument was being conducted on the relative merits of hard-chine and round-bilged hulls for high-speed coastal craft. The *Bold Pioneer,* with a hard-chine design, measuring 120ft overall with 25ft beam, was to be evaluated against the round-bilged *Bold Pathfinder,* 122ft x 21ft beam. The trials, when they took place, provided much valuable information about hull-stressing and construction, but were singularly inconclusive as to the merits of the two hull forms, and the hulls of the successors to the *Bold* class were to be based largely on quite different designs.

This is by way of digression, but it serves to illustrate the position in 1948, when the Admiralty had to find machinery to launch a new MTB/MGB programme. It was decided to use combined diesel and gas turbine machinery for these craft, as ex-German stocks of the well-proven Mercedes-Benz 2500hp diesel were available, and the Admiralty were emboldened by the success of the Gatric trials to commission new gas turbines from Metropolitan-Vickers, based on the Beryl jet engine.

This new marine engine was the G.2, of 4500hp, which went to sea in late 1951 in HMS *Bold Pioneer,* followed in 1953 by her sister ship HMS *Bold Pathfinder.* The machinery arrangement used four separate shafts, with the gas turbines driving on the wing shafts and diesels on the inners. Fixed-pitch propellers were used and no clutches were fitted, the gas turbine shafts being trailed when cruising, and manoeuvring being carried out by direct reversing on the diesels.

The G.2 was a considerable advance on the Gatric, in terms of power, efficiency and specific power (power output per pound of air per second used). The machinery layout was good, and the arrangement proved very successful from an operational point of view. Although neither craft quite achieved its full design speed, both exceeded 42 knots in service, and provided good test-beds for hull construction, propeller design, and other problems encountered in high-speed craft.

The gas turbines, however, ran into considerable teething troubles — and some more fundamental problems — from the start of their operation at sea. The most difficult problem was stalling of the compressor, particularly at low speed, with consequent compressor-blade failures. This was a recurring problem, and although it was alleviated to some extent by modification to the compressor blading and the fitting of blow-off valves, it was aggravated in practice by three other factors — from all of which we could at least learn useful lessons.

Firstly, while stalling occurred normally at low speed, outside the usual running range, the compressor could be brought dangerously near to the stall line at higher speeds by salt deposition on blading, under rough weather conditions. The air intake arrangements were similar to those which had been used successfully — but under much easier conditions — in *MGB 2009,* and again no attempt was made to filter intake air, other than by allowing free water to fall to the bottom of the intake settling chamber. As a result of this, in marginal weather conditions, although the boat herself could maintain 30 knots or more, it could become necessary to wash through compressors with distilled water, to remove salt deposits, at 20-minute intervals, in order to avoid running into compressor stall conditions — a process which was strictly limited by the supply of distilled water onboard.

A second problem arose from the fact that the compressor-blading was of aluminium alloy — a mistake never repeated in subsequent engines — which not only meant that blade vibration, aggravated by salt corrosion, could quickly cause failure, but also usually resulted in an expensive 'haircut' for the whole compressor whenever a blade failure occurred. This problem was made worse by factor number three, a lack of spare parts, which meant that a failure resulted in engines being out of service for long periods while new blades were being manufactured.

Sensitivity to compressor-stall was alleviated by changes in design and could have been further improved. The aluminium compressor blading was the result of adapting an aircraft jet engine at minimum cost, and therefore with minimum change. This was probably justified in the 'Bold' boats, as the G.2 was being used only for two prototype craft, and it could have been modified for later boats of the class. In the event, plans for future fast patrol boats (as they were then designated) were changed, and the G.2 was, in any case, due to be superseded. Metropolitan-Vickers were already, in 1951, working on an modified engine, the G.4, which was of similar size to the G.2 but of more robust construction, with a new compressor using stainless steel blading, and plain journal and thrust bearings in place of the ball and roller bearings. With increased compressor speed, the compression ratio was raised from 4:1 to 6.3:1, with considerable improvement in efficiency and an increase in power to 5000hp.

*The forward engineroom of H.M.S. **Bold Pathfinder**, showing the two G.2 gas turbines, which drove the wing shafts.*

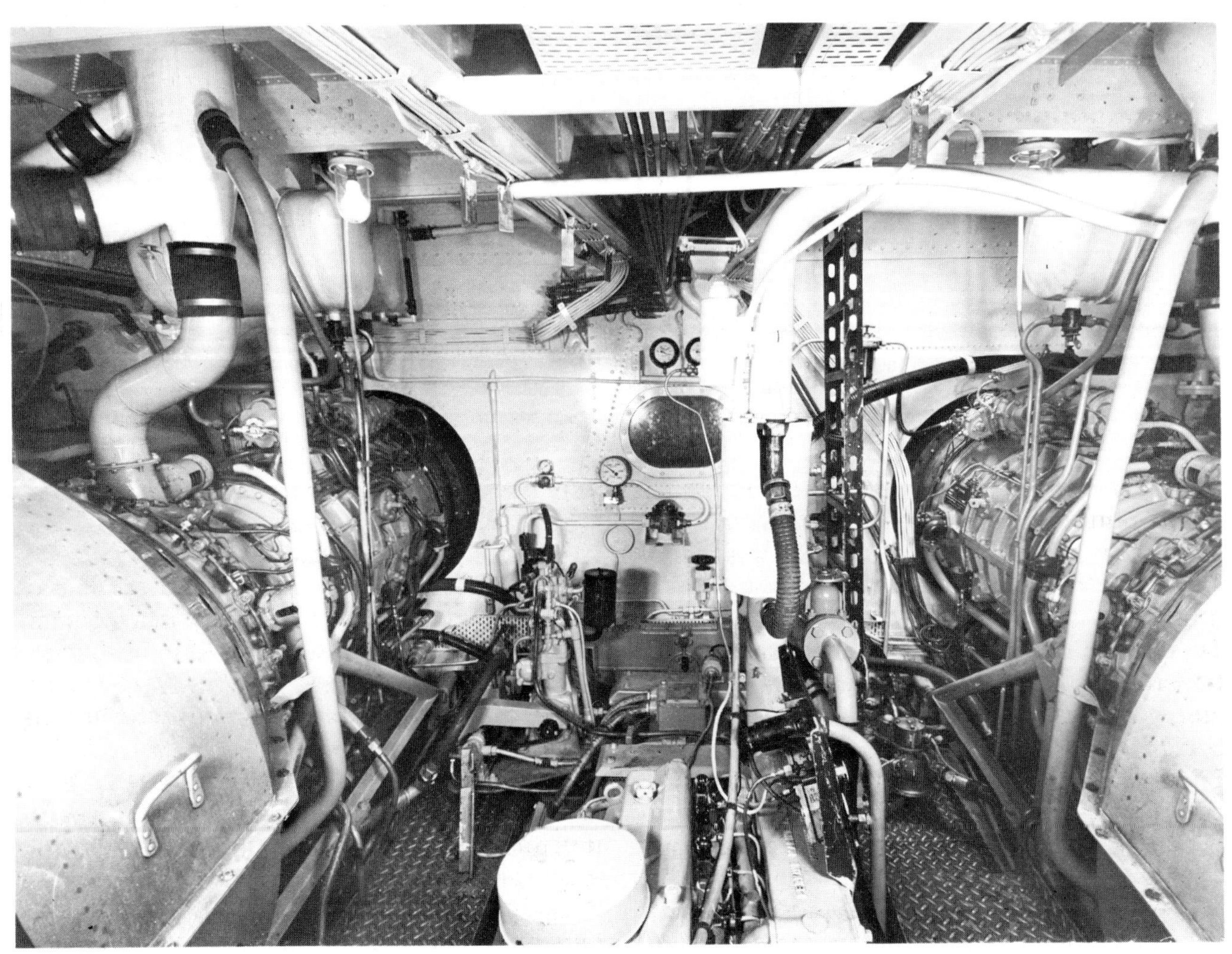

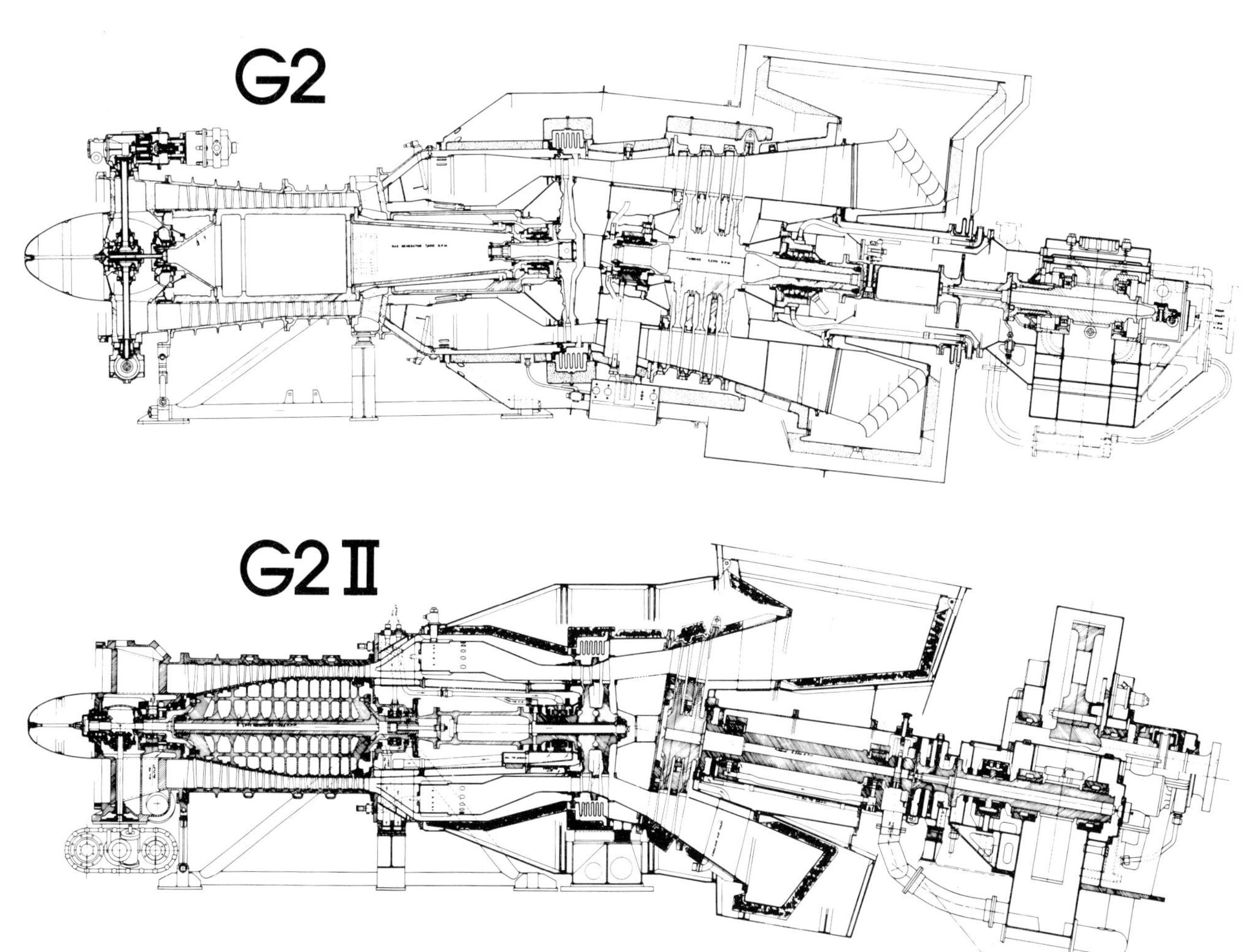

*The G.2 II was an improved version of the original G.2, with a number of minor modifications. The compressor and power turbine shafts were offset 7 degrees, to allow for increased shaft-rake without having to raise engine-room deckhead height. The engine was ordered by the Admiralty for the new class of patrol boats in 1951, but the order was later cancelled and a longer-term plan for development of the G.4 was introduced; but two G.2 IIs were bought by the U.S. Navy and carried out trials in a PT boat. The G.4 was a fully marinized engine, but never went to sea in the Royal Navy, though it was installed in the Italian gunboats, **Lampo** and **Baleno**.*

This new engine was installed in two Italian motor gunboats, the *Lampo* and *Baleno,* both ordered in 1958 from the Italian Navy's yard in Taranto, and it went into service in 1964. It might well have proved a very successful engine in small craft, but for the advent of the Bristol Proteus, which overtook it. It was not destined to be used in the next generation of British fast patrol boats — indeed, changes in British naval policy would soon do away with all operational requirements for fast patrol boats, and would concentrate attention on other types of ship.

It was for frigate and destroyer propulsion that development of the English Electric EL60 A was commissioned by the Admiralty in September 1946. As has already been indicated, the design of this engine was based firmly on steam turbine philosophy, for it was intended to be a long-life unit which could be used in a big ship, either in conjunction with steam plant or for main propulsion. The object was to obtain some preliminary sea-experience with this type of heavyweight engine, hitherto untested, for it was still at that time considered out of the question to utilise a lightweight aero-derived type of unit for big ship propulsion. The ultimate power requirement for a gas turbine in a frigate or destroyer had still to be determined, and would depend on the final machinery 'mix', as well as on ship-operating requirements. The Americans envisaged a combined steam and gas turbine propulsion unit of 30,000hp, in which a high efficiency steam plant would provide 9000hp for cruising — this proportion of the total representing the maximum power requirement of the ship for 99 per cent of its operating time at sea — while gas turbines would provide 21,000hp boost power for the remaining 1 per cent of the running time. These figures may seem exaggerated, but it must be borne in mind that, for a typical 30-knot ship, 30 per cent power represents a speed of something over 20 knots, and the American example assumed that the greater part of the time at sea would be spent in the 15-knot speed range.

It might have been thought that, for the very limited running time required of a boost engine in such an arrangement, a lightweight gas turbine based on aero-engine practice would be acceptable, but questions of shock-resistance and reliability of ball and roller bearings helped to reinforce the cautious instincts of the more conservative marine engineers and, even in boost applications, aero-derivatives were still hardly being considered for big ship installations. In the case of engines designed for cruise propulsion, there was no thought in anyone's mind of using aero-derived units. In those early days, 'lightweight' was still indelibly associated in people's minds with 'short life', so the EL 60A was designed on steam turbine lines.

Certain constraints were imposed by the sea trials plan. It was intended to instal the first unit in HMS *Hotham*, a Lease-Lend ex-US frigate, which the United States had allowed the Admiralty to retain after the war for this special purpose. The *Hotham* was originally fitted steam turbo-electric machinery developed 6000 hp on each of two shafts. She could readily be converted for use as a floating gas turbine testbed, since reversing was already available in the electrical transmission system, and if a gas turbine of suitable size were to replace one steam set, a direct comparison could be made between gas turbine and steam propulsion.

The EL 60A was therefore designed as a 6500hp unit, and the layout was adapted to fit in existing machinery spaces in the *Hotham*. Since the requirement was for a main propulsion or cruise role, the designers adopted a regenera-

tive cycle, with the exhaust gas being used to heat the air from the compressor before its introduction into the combustion chamber, in order to provide reasonably good efficiency, particularly at part-load. The engine was designed on very conservative lines, with a low compression-ratio and very modest turbine temperatures; but it was at the expense of accepting a large, heavy unit, which swallowed an enormous quality of air, thus requiring a large amount of space for air intake and exhaust ducting.

Almost at the same time as the award of the contract for the EL 60A, the Admiralty ordered a gas turbine of an entirely different stamp. This was the Rolls-Royce RM 60 an engine which was to mark a further milestone in progress when it went to sea in HMS *Grey Goose* in 1953, for she was the world's first warship to be propelled soley by gas turbines.

The original intention in designing the RM 60 had been to produce an engine for coastal craft which had good efficiency at low power, in order to provide economical cruising as well as high-speed propulsion. By the time that it entered service, it was already becoming recognised that the complex cycle of the RM 60 introduced a degree of sophistication which went beyond the requirements of the average fast patrol boat, and that such a unit might be considered for the role of cruise engine in larger ships. As time passed, it became clear that the right solution for the small high-speed craft was a simple-cycle gas turbine, giving maximum economy in weight and space with reasonable efficiency at high speed; and that if long endurance at low speed were required, this could best be achieved by running a single gas turbine in a multi-engined installation, or, more dramatically, by using small cruising diesel engines, which gave a huge increase in endurance at the expense of a very low cruising speed.

At the same time, the trials of the RM 60 did demonstrate to a doubting marine world that 'lightweight' did not necessarily connote 'short life'. The *Grey Goose* continued in service for over four years and operated extensively in the UK waters and in the Mediteranean. Her engines built up a total of 1400 running hours, and they proved remarkably reliable, despite their considerable complexity. No doubt this was due in part to the quite different development

The Rolls-Royce RM 60 gas turbine, 5400 hp. A great step forward in marine gas turbine technology, but too large and complicated for its intended role as a fast patrol boat engine.

*H.M.S. **Grey Goose**. Built as a steam gunboat, and commanded during the war by Peter Scott, she was converted to carry the RM 60 in 1953, and ran for three years as a highly successful trials boat.*

philosophy adopted by Rolls-Royce, and the long shore-testing period which the engines underwent before installation, involving a large number of hours run on the test-bed, by the standards of those days.

By way of comparison, the first G.2. engine, following the success of the Gatric, was run for only 37 hours on the test-bed before carrying out an official nine-hour acceptance test, prior to installation in *Bold pioneer* . The RM 60 was run for 1100 hours as a complete unit, after 640 -

hours initial testing of the high-pressure section, before installation in the *Grey Goose*.

It was of course only natural that the RM 60 should require much more comprehensive development testing than the contemporary G.2., because, in the first place, it was not based on any existing aero engine — even though its design was certainly based on aero-engine technology — and, secondly, it used an extremely complex cycle, with a very high compression-ratio. This was 18 : 1, compared with only 4 : 1 in both the G.2. and the EL 60A — giving an internal pressure in the combustion section, and at entry to the turbine, of over 270lbs/sq inch, as opposed to 60lbs/sq inch. Such high pressures had never before been used in gas turbines, and were not to be reached again until the introduction of more advanced jet engines in the 1960s. They raised new problems in combustion-chamber design and sealing, and, operating at relatively low maximum temperatures, they involved use of two stages of intercooling in the compressor, as well as a regenerative heat-exchanger to give improved part-load fuel economy.

Despite these complications the RM 60 went through its period of development and shore testing without serious trouble, except for problems with ball-bearings, which necessitated a number of modifications. At the time this rather shook the confidence of marine engineers in this type of bearing for 'serious' naval applications.

Meanwhile, the EL 60A had encountered problems in development which were to a large extent inherent in the overall problem of adapting steam technology, both in design and in manufacturing processes, to the requirements of the new type of prime mover. By the autumn of 1951, the EL 60A was finally ready for full-scale shore trials; but by this time the RM 60 had completed its initial test bed running and plans were well advanced for putting it to sea. It was rapidly becoming realized that the original approach to the question of big ship gas turbines, through the development of heavy and complex 'steam-derived' units, was mistaken. In consequence, the sea trials in HMS *Hotham* were cancelled, and the EL 60A programme was terminated in 1952, on conclusion of the initial shore testing.

The engineroom in H.M.S. **Grey Goose**. *The RM 60s provided the sole means of propulsion, driving through two Rotol C.P. propellers.*

ACCEPTANCE ~ FOR SMALL CRAFT

The gas turbines which we have been reviewing — in fact all gas turbines in naval service up till 1958 — were essentially experimental engines, designed to explore the potential of the new type of prime-mover and to gain sea experience, but never intended to be used as production engines in a class of new ships or fast patrol boats.

The G.2. is no exception to this rule; although the experimenting in this case was with the boats rather than with the engines, the latter were installed as a quickly available stop-gap, and Metropolitan-Vickers' candidate engine for the production boats was the G.4.

The year 1958 marks a significant step in the advance of marine gas turbines, as it was in this year that the first full-scale production engine, the Bristol 'Proteus', started sea trials in HMS *Brave Borderer*. The Proteus was a rather remarkable engine, because for some time it had virtually no challenger, and although improved diesels and new gas turbines have been developed, it has nevertheless remained in production without intermission for more than 20 years, being used in no fewer than 16 different types of patrol boat, hydrofoil and hovercraft, and seeing service in 12 of the world's navies, as well as in the familiar cross-channel hovercraft ferries.

As an aero engine, it was used as the power plant of the Bristol Britannia, in its day the largest and most advanced of the post-war passenger aircraft, and the first turbo-prop airliner on the trans-Atlantic service. But the story of the Proteus really dates from 1941, when Bristol Aero-Engines first embarked on gas turbine development. The initial plan was for a propeller turbine of 4000hp, with a specific fuel consumption at least as good as contemporary piston engines, to power an aircraft capable of 300 knots at a height of 20,000 feet. The design was later scaled down to 2500hp, and the engine, known as the Theseus, was the first turbo-prop to complete the Air Ministry 100-hour test, in December 1946.

In early 1947 the Theseus started flight-testing, with two units installed in a Lincoln bomber, but by this time a more powerful engine was already being developed, based on the experience gained with the Theseus, for use in two new aircraft projects, the Bristol Brabazon and the Saunders-Roe Princess flying boat, and on 25th January, 1947, the Proteus ran for the first time on the test bed.

One requirement of the two new aircraft was for a very short engine, but room was still needed for a good length of combustion chamber. Bristol tackled this problem in a highly ingenious fashion, by turning the compressor back-to-front, so that the air was drawn in to the mid-section of the engine, passed forward through the compressor, turned through 180°, and then passed aft through the combustion chambers, which were grouped around the outer casing of the compressor. The result was an exceptionally short, stocky unit, very rigid, and with virtually the whole weight of the engine carried on a single mounting ring. This unusual configuration was to make the engine particularly suitable for installation in boats when it came to be considered for a marine role later on.

The Brabazon proved unsuccessful, and the Proteus-powered version of the aircraft was never built; but three Princess flying boats were completed, each powered by an awe-inspiring array of ten Proteus, made up of one single and two twin units in each wing. And even though these engines were purely aero-units, it was appreciated that they would have to put up with a good deal of salt water, so that, very early in its life, the Proteus had its first taste of sea conditions, through being hosed down with water on the test bed. This was fresh water: when Bristol wanted salt water later on for marine engines, they tried importing sea water at Avonmouth from a coaster, but got into trouble with the Customs! Thereafter, 'standard' sea water was made up artificially — a necessary method for proper testing, since the real thing varies greatly in different parts of the world.

After many vicissitudes the Princess project was abandoned. It seems sad that these elegant and ambitious flying boats were never put into commercial service, and had to lie rotting at their moorings, a monument to progress, until they were finally scrapped in 1965. But although the first Princess flew successfully on 22nd August 1952, it had already, in effect, been superseded just a week earlier, when the first Britannia made its maiden flight, powered by four Proteus. A considerable interval was to elapse before the Britannia entered passenger service in 1957, and further development work was being done on the Proteus bringing the power up from 3500hp to 4400hp; but during this period of in-flight development, new plans were being unfolded in the Royal Navy.

There was much debate in the early 1950s over the role of Coastal Forces and the type of Fast Patrol Boat which the RN would require for the future. A number of 'long boats' — Fairmile 'D' torpedo boats and Camper & Nicholson gunboats — were still in service, and the Korean War brought a flurry of new activity. The 'Gay' class, introduced in 1952, were, in effect, copies of late wartime 'short' boats powered by three Packard petrol engines. These were followed in 1954 by the 'Dark' class, stouter craft of composite build, and powered by the new 2500hp Napier Deltic diesel, but still 'short' boats, of 70ft length, and limited in their armament. The two torpedoes were changed from 18inch to 21inch, but gun armament was confined to a Bofors 40mm or the rather comic Coastal Forces 4.5inch 8cwt pop-gun. The boats had been designed originally for 45 knots, but the displacement had increased during the design-

phase, and they were limited to just under 40 knots in service.

At the other end of the scale were the two 'Bold' class boats, better sea boats, and better equipped for independent operation, but unnecessarily large for the task which they would be required to perform and for the weapons which they were intended to mount. Provided craft could be made truly convertible from gunboats to torpedo boats, in a matter of hours, a significantly smaller hull could be used, with economy in cost, improved performance, and better manoeuvrability.

In 1954, a new Staff Requirement for Medium Fast Patrol Boats was drawn up, and Vosper were asked to undertake a design-study. The aim was to produce a boat of about 100ft overall length, with a speed of 50 knots and an endurance at maximum continuous speed of 400 miles. It would carry four 21inch torpedoes in the MTB role, and as a gunboat it would mount the new CFS2 3.3inch gun. This latter was something of a novelty in Coastal Forces weaponry, developed from the Army's 17pounder tank gun, with a fully stabilized mounting. The result was very heavy, and added to the problems of the boat-designer, but there is no doubt that the whole concept of the new fast patrol boats was far in advance of anything that had been used in Coastal Forces hitherto.

Vosper studied a number of machinery operations. These included originally three Napier 'Deltic' diesels, of 2500hp each; two Metropolitan-Vickers G.4 gas turbines, of 5000hp each; two Deltics and one G.4; and two Compound Deltics developing 5000hp each. During the early discussion of the project, it was decided that a further three-engined arrangement should also be considered, using the Bristol Proteus, which was then well advanced in its flight-development programme, and which had been put forward for marine use at 3500hp. The three-Deltic arrangement was eliminated at a relatively early stage in the study as, within the limitations imposed by the armament and range requirement, it could not achieve more than 40 knots. The Compound Deltic was an ingenious combination of the advantages (or, as some said, disadvantages) of both the diesel and the gas turbine. It incorporated within the triangle of pistons and cylinders a turbo-blower, which supercharged the diesel and also

The Marine Proteus, 4500 hp. The Proteus required little modification to fit it for marine use, other than changes in material and in the fuel system, but a new primary reduction gear was produced by W.H. Allen, to give a higher output shaft speed, suitable for use with a V-drive reversing gear.

contributed some 30 per cent of the output shaft power. Such an engine would meet the power requirement, but would need time and money for development, and there were doubts as to whether a complex and relatively untried engine system would prove suitable for the rough-and-tumble life of a fast patrol boat.

Both diesel arrangements also suffered from two other fundamental disadvantages. In the first place the piston engine's basic power-speed curve is ill-matched to the corresponding hull-requirement curve for a high-speed planing hull. This means either sacrificing top speed when the boat is running light with a clean bottom, or risking 'locking-up' of the diesels, at disastrously low speeds, in the 'deep-and-dirty' condition. Secondly, the diesel cannot operate satisfactorily at low power, except for short periods. This would have meant, in the three-Deltic version, that the minimum speed on patrol, even using only one engine, would have been some 10 - 12 knots, and in the twin-engined boat with the compound Deltic the minimum speed would have been much higher.

It was clear, then, that for the task which was required, only gas turbines could fully meet the

requirements. The G.4. remained a possibility, but more detailed study showed that the three-Proteus arrangement could provide a number of advantages, including reduced fuel and machinery weight, greater range, higher speed, better low-speed endurance, and a more practical disposition of machinery within the hull. In fact, the Proteus-powered design would meet the full staff requirement for 50 knots maximum speed and a 400-mile endurance at 43 knots, on a 98ft hull with a deep displacement of 114 tons.

After nearly two years of debate and experiment, Vosper received the contract in March 1956 for the construction of two Medium Fast Patrol Boats, and the Proteus was launched on its naval career. But satisfaction at the outcome of the long debate was to be short-lived, for the whole future of the high-performance gas turbine-powered fast patrol boat was to receive a serious set-back before the first of the 'Braves' (as the new class were christened) had even been completed.

In 1956 the British Government embarked on a re-appraisal of British naval commitments, and as a result of this it was decided that, within the NATO plan, the Royal Navy's effort should

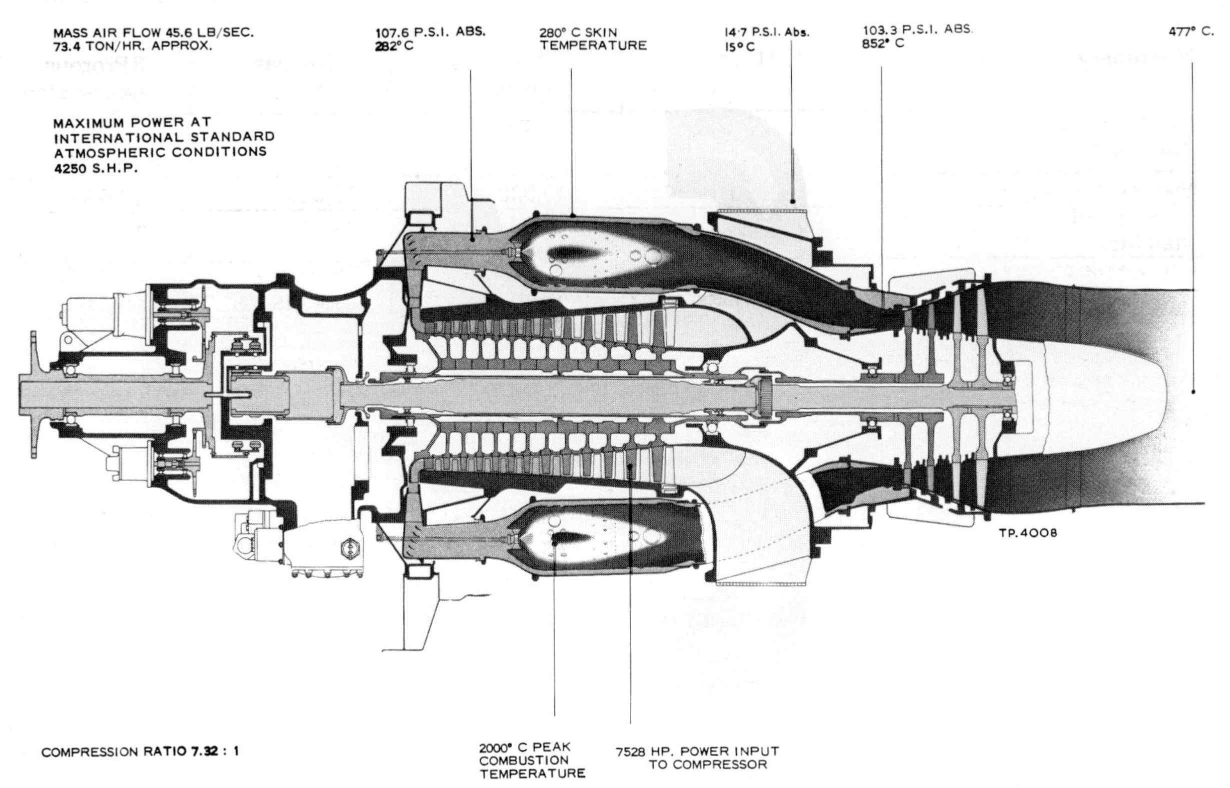

The Proteus's unusual reverse-flow arrangement produced a short, compact, and rigid engine which could be supported on a simple three-point mounting, and had the added advantage, from a boat installation point of view, that the drive was taken from the "cold" end.

be concentrated in the Atlantic theatre, leaving inshore defence in the North Sea and Baltic to other NATO allies. We are obviously not concerned here with the strategic implications of this decision; in practical terms, however, it meant virtually the complete abandonment of Britain's Coastal Forces, the paying off of existing operational boats, and a stop to all new building. The Admiralty seemed to be repeating what had happened in 1922 when, once before, Coastal Forces had been disbanded for 14 years until, in 1936, a flurry of ill-coordinated activity had led the RN into the Second World War with a pitiful collection of badly-armed and inadequate MTBs. However, when the axe fell in 1957, the first two prototype boats of the new class were already well advanced, and it was agreed that they should be completed and commissioned for trials, although the new Coastal Forces gun was abandoned and the two boats were never destined to form an operational unit.

Nevertheless, the fact that they were completed, and did commission in the RN, made it possible to demonstrate the potential both of the hull and of the new gas turbine. Over the next 14 years, a total of 23 craft were to be built to the basic Vosper high-speed hull design and powered by Proteus engines, and they were to see service in the navies of Germany, Denmark, Greece, Libya, Malaysia, and Brunei, as well as in successive British applications. It is, perhaps, a measure of the success of the engine, and certainly a graphic illustration of one advantage of the gas turbine, that when in 1970, after 12 years' service — a reasonable 'expectation of life' for a fast patrol boat — the two 'Brave' boats were finally scrapped and replaced by three twin-engined 'Fast Training Boats', the original Proteus engines were removed from the old boats, overhauled, uprated to 4500 hp, and installed in the new boats, where they remain in service today.

Although the Proteus has also been used in a

"Brave" Class Fast Patrol Boats

Estimated performance, assuming similar hull dimensions and weight, and identical equipment and armament:

Machinery	3 Deltic diesels	2 Compound Deltics	2 G.4. gas turbines	3 Proteus gas turbines
Displacement (half fuel)	102 tons	99 tons	106 tons	102 tons
Max. power, s.h.p.	7500	10,000	10,000	10,500
Max. speed (half fuel)	40 kts.	45 kts.	44 kts.	49 kts.
Endurance at max. continuous speed	424 miles at 31 kts.	372 miles at 34 kts.	366 miles at 36 kts.	404 miles at 43 kts.

Brave machinery comparison. A large number of variants were examined, but the above figures show the best which could be obtained with the four basic machinery alternatives. The three-shaft arrangement, with high-speed fully cavitating propellers, reduced shaft rake and made a considerable improvement in propulsive efficiency.

variety of conventional patrol boats — in Sweden, Denmark, Italy and Yugoslavia — it was particularly associated in the early days with the 50-knot hard-chine hull which sprang from the genius of Peter du Cane, and made such a spectacular entrance on the naval scene in the 1960s. But the career of this new breed of craft was to be a limited one, and the type was never to reach its full potential for a number of reasons. In the first place, none of the boats built was ever equipped with the armament which had originally been intended, the CFS2 gun. Since the RN no longer had any coastal forces role there was no move to develop any effective new weapon, in the form of a lightweight missile system, even though the Egyptians, in the 1967 war with Israel, had demonstrated the possibilities of missile-carrying patrol boats when they sank the Israeli destroyer *Eilat*. A second reason for the premature decline in the hard-chine boat's popularity was that the navies most closely concerned with coastal warfare, and most influential in the development of coastal forces doctrine, were the Baltic navies — German, Danish, and Swedish — all of whom had similar ideas on the role and function of fast patrol boats, which depended on their ability to operate in the short, steep Baltic seas.

A lot of rubbish is talked about 'special' weather conditions which many people claim to encounter in their own particular part of the world, and although it is common to speak of the 'shallow waters' of the Baltic, the North Sea and Channel are, in fact, shallower, and can certainly produce equally nasty conditions. Nevertheless, there is undoubtedly a case to be

made for the larger, heavier, round-bilged boat, under either Baltic or North Sea conditions, which is perhaps based more on the need for reasonable habitability and the capacity for independent operation, rather than on the sheer physical capacity to cope with rough weather. If a bigger hull were needed, and boats were to increase in size from the 100 tons of the 'Brave' class to the 200 tons or more of the later Ger-

One disadvantage of the diesel in a high-speed planing hull. The power required to drive the hull, plotted against speed, follows a similar line to that of the diesel's maximum power v. speed curve. If, through an increase in displacement and/or fouling of the bottom, the hull power requirement is increased, then the diesel can "lock up" at reduced speed and correspondingly diminshed power. The gas turbine, on the other hand, can still produce almost maximum power, even when the speed of the power turbine is cut back by boat or propeller limitations, so the effect of increased hull resistance is minimized.

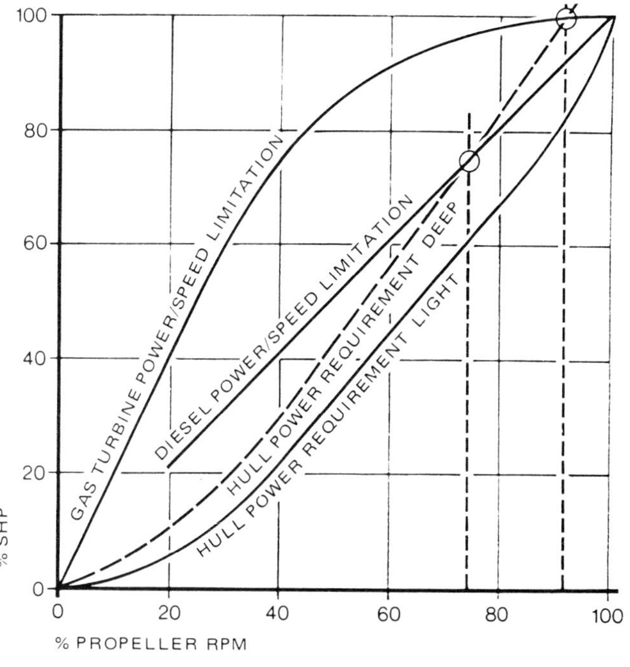

man, Swedish and Danish craft, then maintenance of the 50-knot capability, whatever the hull form, would need a large increase in engine power, whereas, even in the larger hull, the three-Proteus arrangement could still provide a respectable 40 knots. At this speed, it may be worth noting, even the larger boat would still, in practice, be 'planing', and all the round-bilged hulls used a flat-bottom form in their after sections; but it would gain little in propulsive efficiency by going to the extreme wide-beamed hard-chine form — as had already been demonstrated with the trials of the 'Bold' class.

The Germans had developed a breed of E-Boat which had had some considerable success during the war, and had shown itself technically superior to its British counterpart. The E-boat was based on a narrow, round-bilged hull of sturdy, steel-framed, composite construction, mounting little gun armament and relying on torpedoes. It had a speed of 40 knots, was inclined to be wet in head-seas, but was able to maintain speed to windward in rough weather conditions more comfortably than the average hard-chine boat. With a Navy and a boatbuilding business weaned on the E-Boat type of hull, it was perhaps not surprising that the hard-chine boat should have had limited appeal, and although two prototype boats, *Pfeil* and *Strahl* were built by Vospers for the Germans in 1962, there was never any great enthusiasm for them in naval circles. It was accepted that, for all-weather operation in the Baltic, the best compromise was a larger, more robust, round-bilged hull, ten knots slower than the Vosper boat in fair weather, but able to out-sail her in more marginal conditions, and with greater crew comfort and autonomy.

When the Royal Swedish Navy started to plan a new class of MTB in 1959, they made a detailed evaluation of alternative hull designs, including the Vosper 'Brave' hull and a German patrol boat of Lürssen design, and carried out exhaustive tank tests. Their conclusion was that, for their Baltic role, the Lürssen hull, developed from war-time E-boat experience, was the best solution.

The Swedes also made a detailed evaluation of alternative machinery arrangements. The final choice lay between four Daimler-Benz MB 518B diesels, each developing 3000 hp (metric), and three Bristol Proteus gas turbines rated at 4310 hp (metric). The Navy concluded that the gas turbine arrangement, though it had a higher fuel-consumption than the diesel equivalent, had the advantage in total fuel-plus-machinery weight for any range up to 700 miles — more than sufficient for the Staff Requirement — as well as in quick starting, continuous high-speed operation, and total running costs. The gas turbines also effected a saving in machinery space of 20 per cent and enabled the complete propulsion machinery to be located right aft, in a less valuable section of the hull. Perhaps the most interesting comparison, in view of later developments in the use of gas turbines, was in maintenance, availability, and manning. The Swedes concluded that onboard maintenance with gas turbines would amount to less than one-thirtieth of that required by the diesels, and that, when engine change was required, it could be done in one-tenth of the time; the diesel-powered boat would require an engine room complement of twelve, the gas turbine only four.

In 1966-67 the Swedish Navy commissioned six *Spica* class three-Proteus boats, and these were followed in 1973-76 by a further 12 boats of similar type. The Royal Danish Navy, which had originally built six Vosper-type hard-chine boats in 1965-66, also decided to move to the round-bilged type of boat, and commissioned ten much larger craft of 260 tons displacement, but still retaining the three-Proteus layout, between 1976 and 1978. These *Willemoes* class boats marked a further advance in the development of the patrol boat role for, like their contemporaries in France and Germany, they were armed with guided missiles as well as torpedoes. They also used a machinery arrangement slightly different from the British and Swedish boats, in that they incorporated small diesels on the wing shafts, giving them a modest cruising speed of 12 knots, but with an endurance measured in days rather than hours.

This was, of course, a CODOG* system, with the diesels running independently of the gas turbines, though driving through a common gear box to the same wing propellers. During the same period of the early 1960s, however, the Italian Navy were building a class of fast patrol boat which used true combined diesel and gas turbine machinery — a CODAG system, with two large diesels providing a total of 7600

See Glossary, p. 112.

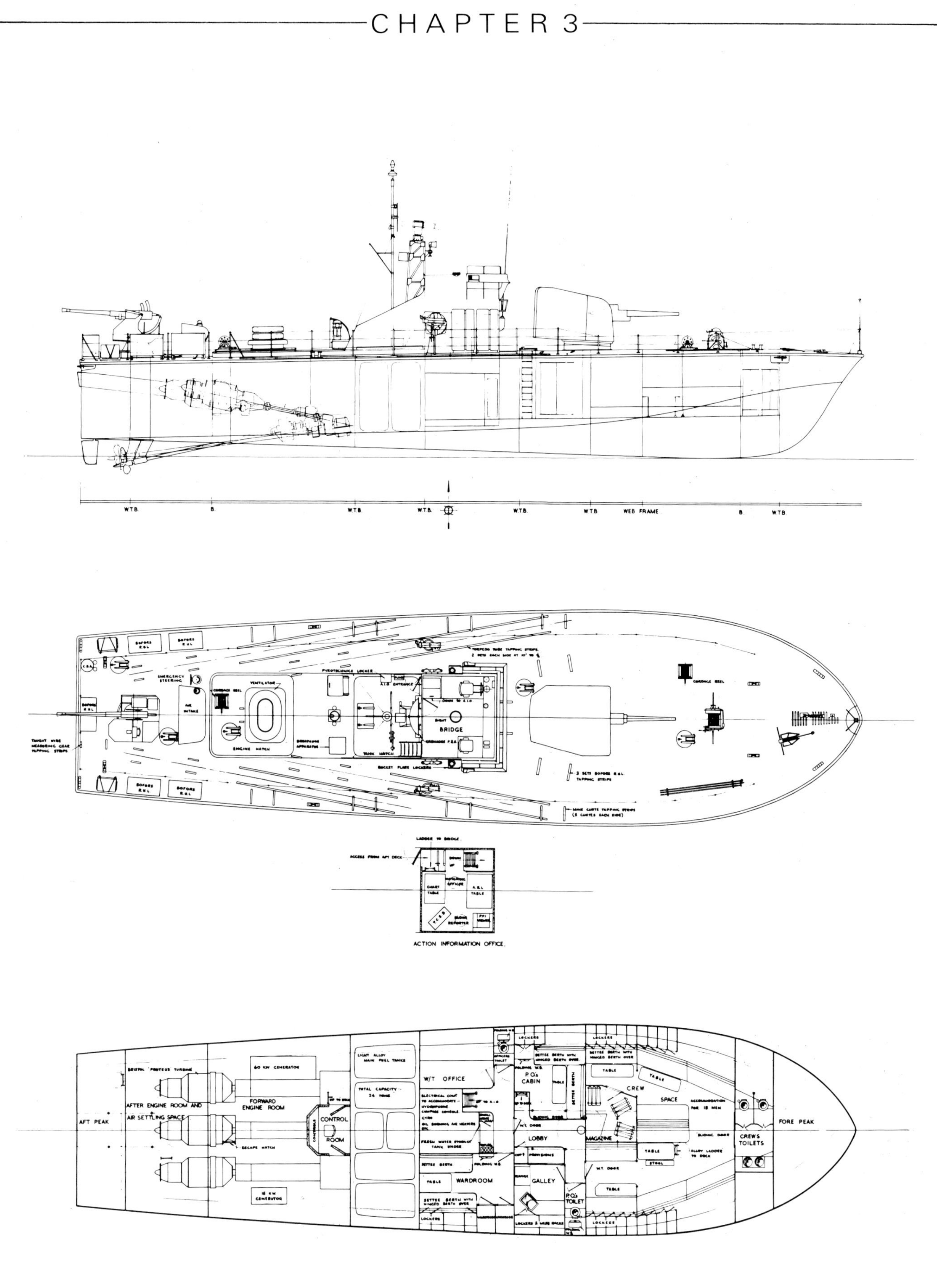

The final arrangement of the 'Brave' class fast patrol boats. The bridge was placed unusually far aft, to accommodate the 9 ton 3.3 inch gun turret. The use of a V-drive transmission enabled the engines to be placed right aft, with a short exhaust through the transom, and fuel tanks could be located amidships, so that trim was not affected as fuel was used up at sea.

hp to give a speed of 30 knots, and a single gas turbine driving on a separate centre-line shaft to boost maximum speed to 40 knots. The first two boats, *Lampo* and *Baleno,* were originally fitted with the Metropolitan-Vickers G.4. gas turbine; the *Freccia* and *Saetta,* completed in 1965-66, used the Proteus, and *Lampo* and *Baleno* were also converted later, in 1968 and 1973, to take the Proteus. These craft were originally desig-nated gunboats, though they were all convertible to the torpedo boat role. The *Saetta* was later equipped with Contraves Sea Killer missiles; the other boats were never re-armed, but they were to be followed by a much more formidable craft of a new type, mounting the Otomelara 76mm gun and the Otomat missile system. This was the *Sparviero* class hydrofoil, powered by a single Proteus gas turbine, now uprated to a maximum of 5000 hp, to provide extra power for 'take-off'.

In man's attempts to escape from the limitations of the normal displacement craft, the hydrofoil has received much less attention than the planing hull, presumably because it is less obvious and more difficult; but it can pay big dividends in improved high-speed performance. The Germans did a considerable amount of experimental work in this field before the war, and this was continued later in Lucerne by the Supramar company, which fathered a whole generation of small passenger hydrofoils, the larger number built in Italy by Cantieri Rodriguez (now Navaltecnica) of Messina. In the military sphere, both the Americans and the Canadians experimented with hydrofoils during the post-war years. The trouble with all these earlier hydrofoils, however, was that they employed a fixed, surface-piercing foil system, unable to adapt to wave contours, which tended to make for a very bumpy ride, particularly in short seas, and virtually confined the craft to fair weather operation or sheltered waters, so that they had very limited military potential.

The first successful step to remedy this shortcoming seems to have been the building by Boeing, under contract to the US Navy, of the USS *High Point,* which commissioned in 1963. This craft, powered by two of the ubiquitous Proteus gas turbines, used a fully submerged computer-controlled foil system, which took account automatically of wave height, vertical acceleration, roll and pitch, to provide steady flight at a constant pre-set height and an extremely stable platform. The fully submerged foil system not only extended the passenger-carrying potential — at considerable increase in cost — but also made it practicable to apply hydrofoils to military purposes by rendering them much less dependent on weather conditions. Following the successful running of *High Point,* two new prototype craft were commissioned in 1968 for comparative evaluation, each using a slightly different foil configuration, and a different system of propulsion. These were the Grumman *Flagstaff,* powered by the Rolls-Royce Tyne, and Boeing *'Tucumcari'* which used what we must now refer to as the Rolls-Royce Proteus*.

The *Flagstaff,* or rather the civil version, which was built two years earlier by Blohm and

---

**It should be noted that, while British naval gas turbines have originated in a number of different companies, the majority of the propulsion engines (as opposed to auxiliary gas turbines) now come under the aegis of Rolls-Royce, through the merging of most of the main gas turbine builders.*

1959 Bristol Aero-Engines and
Armstrong Siddeley Motors merged to
form Bristol Siddeley.
1961 De Havilland Engines and Blackburn
Engines joined Bristol Siddeley.
1967 Bristol Siddeley merged with Rolls-Royce.

Rolls-Royce marine engines, or their aero progenitors, originated as follows:-

Bristol Aero-Engines .... Proteus, Olympus
Armstrong Siddeley
*Motors .... Viper**
De Havilland Engines
*(with General Electric) .... Gnome**
Blackburn Engines
(with Turbomeca) .... Turmo,*
Nimbus,*
*Marbore**
Rolls-Royce .... RM 60, Tyne,
Spey, RB 211
Metropolitan Vickers, who produced the Gatric, G.2., G.4., and G.6., and English Electric, who built the EL 60 A, now both form part of G.E.C.
**Hovercraft engines, qv, Chapter 5.*

Voss, under licence from Grumman, was originally to have been powered by the Proteus, but a last-minute decision was made to substitute the Tyne (it should be remembered that, at the time, Bristol Siddeley and Rolls-Royce were competitors) and an interim lightweight version of the engine was used, based closely on the aero-engine configuration, with only minor changes in materials and in the design of the air intake duct. Much more radical changes were to be made when the engine was fully marinized, as a long-life cruise engine, during the 1970s, but this is a subject for another chapter (vide Ch. 7).

The *Tucumcari* used a similar foil arrangement to *High Point,* but propulsion was by means of a water jet. This was operated by a pump, coupled directly to the Proteus output shaft, which drew water up through the twin after foils and ejected it through twin nozzles under the transom. Although, even at high speeds, this form of propulsion entails some loss in efficiency — the water jet only shows a gain over a good super-cavitating propeller at speeds above about 65 knots, beyond the reach of *Tucumcari* — there is a major gain in simplicity, with all the complication of long propeller

shafts or bevel gearing avoided. Furthermore, since *Tucumcari's* three foils were all retractable, the absence of a propellor meant that, when she was cruising in the hull-borne condition, she had very shallow draught, with virtually no underwater appendages beyond the small retractable propellor which was used for cruise-propulsion.

Tucumcari and *Flagstaff* continued their evaluation for some years, and *Tucumcari* in particular proved very successful; but her career was, alas, cut short dramatically in 1973, when she inadvertently ran aground on a coral reef in the Virgin Islands. It is perhaps some small consolation that, although the craft was wrecked, her Proteus engine was unharmed, and continued in service as a spare unit for *High Point.* More to the point, *Tucumcari* was the prototype for the Italian *Sparviero* class.

A large craft, using the same foil and jet propulsion system, though with Allison engines, was later developed by Boeing for passenger service, and one of this type, HMS *Speedy,* was commissioned in the RN in 1980, to evaluate the possibilities of hydrofoil operation in the North Sea.

The Proteus has managed to put up with a

H.M.S. **Brave Borderer** *and* **Brave Swordsman** *in close formation at 50 knots. Fine boats that never reached their full potential.*

*The Royal Navy's fast target boats, **Cutlass**, **Scimitar**, and **Sabre**. The **Brave** class boats were finally scrapped in 1971, but their original Proteus engines were overhauled, uprated, and installed in the new **Cutlass** class, which were of similar hull design.*

*The **Spica** class engine-room layout is similar to that in the Vosper hard-chine boats, but reversing is by KaMeWa C.P. propellers. For maximum cruising endurance, boats can run on the centre engine with the wing propellers feathered.*

*The Royal Danish Navy's **Willemoes**, 260 tons. With these boats, the Danes adopted a Lürssen hull design, similar to the Swedish **Spica** class, but retained the same Proteus CODOG machinery as their earlier **Søløven** class. Later boats carried Harpoon missiles aft.*

good deal of rough treatment during its career, including not only wrecking on coral reefs but hard slamming in head seas, swallowing heavy spray, ingesting sand in hovercraft and running (in its industrial version) on heavy fuel. A story is told that, during the early trials of *Brave Borderer,* the crew filled the distilled water washing tank from a carboy on the jetty before going to sea, and duly washed through the compressors of the engines as they returned up harbour in the evening. As they neared the jetty, they were hailed by a gesticulating figure, asking if they had washed compressors.

'Yes, of course', they assured him.

'Well, now you really have fouled it up,' he replied. 'That carboy you filled up from this morning — that wasn't distilled water; it was sulphuric acid!'

The engines were hastily removed from the boat, stripped, cleaned, and examined; but they proved to be in sparkling condition and none the worse for this rather drastic treatment. The Royal Navy, however, has not yet adopted sulphuric acid as a regular cleaning agent.

Another story, which has nothing to do with engine reliability, concerns the first industrial Proteus, which was installed in a small unmanned generating station in Princetown, on Dartmoor. The set was remotely controlled and monitored from Bristol, 100 miles away, using the ordinary GPO telephone system, and engine controls included an automatic dialling device with a taped voice call-up, by means of which the plant could ring Bristol and request instructions in the event of a fault developing. The set was started one Friday and inadvertently left running, with the result that on Sunday it ran out of fuel. It duly shut itself down and called Bristol. Unfortunately, the GPO's telephone engineers had been busy in the interval, with the result that, throughout the whole of Sunday night, before a routine check on Monday morning, two robots — a man in Princetown and a lady in Bristol — carried on an ex-

---

"Spica" Class Motor Torpedo Boats
Diesel versus gas turbine comparison

Machinery	Gas turbines (3 Proteus)	Diesels (4 MB 518 B)
Max. installed power, metric b.h.p.	12,930	12,000
Total machinery weight	20 tons	38 tons
Engine room crew (including equipment, provisions, and water)	3 tons	9 tons
Fuel for 400 miles at max. continuous speed	29 tons	17 tons
Total fuel-plus-machinery weight for 400 miles endurance:		
Deep displacement	52 tons	64 tons
½-fuel displacement	37.5 tons	55.5 tons
Fuel, maintenance, and overhaul cost, per hour run (Swedish Kroner)		
Total	Kr. 629	Kr. 749
Onboard maintenance	Kr. 5*	Kr. 180

**This was an early estimate, based on limited operation in the 'Brave' class fast patrol boats and the Swedish T.101 MTB which was used for Proteus trials in the Baltic. In practice, the figure has probably been higher, but the difference between gas turbine and diesel is still striking.*

The Swedish Navy had had some earlier experience with gas turbines, in a 60 foot patrol boat which was equipped with a De Havilland 'Goblin' engine in 1950. They also carried out sea trials with the Proteus in an existing M.T.B., T.101, before making the final decision on the Spica class design; but all other boats in service at that time were diesel powered.

tremely boring but unfailingly good-natured conversation:

'This is Princetown calling Bristol. Request instructions.'
'This number has been changed. Please dial Bristol 72509.'
'This is Princetown calling Bristol. Request instructions.'
'This number has been changed... '

Various attempts have been made to replace the Proteus. As early as 1965 Bristol Siddeley decided that, with the progress that had been made in gas turbine design, and the increase which was already apparent in the displacement, and therefore power requirement, of new fast patrol boats, there would in the future be a need for a new gas turbine, of some 6000 to 8000hp, which could have an improved fuel consumption. Such an engine would have applications not only in the new breed of fast patrol boat, but also in larger hovercraft and hydrofoils which were then being foreseen.

There were a number of possibilities. Probably the most obvious was the Rolls-Royce Tyne, but this was before the days of the Bristol Siddeley — Rolls-Royce merger, and the intitiative to replace the Proteus was a purely Bristol Siddeley one. There was, at that time, no interest in a Proteus replacement on the part of the Admiralty. The Tyne, in any case, would provide little increase in power, though it did have a greatly improved efficiency, and it was not until 1978 that an uprated version of the engine was to be introduced into service, capable of reaching 6000hp.

*The Italian Navy's motor gunboat **Freccia,** built in the mid-1960s, with Proteus/Fiat diesel CODAG machinery.*

*The Italian Navy's **Sparviero,** first of a class of seven hydrofoils, water-jet propelled by a single Proteus, and mounting a formidable armament for a vessel of this size.*

*U.S.S. **Flagstaff**, one of two prototype hydrofoils, built by Grumman for the United States Navy, with a single Tyne and a C.P. propeller.*

A more promising candidate, though still in the early stages of development, was a new turbo-fan engine, the M.45, which was being developed jointly by Bristol Siddeley and the French firm SNECMA, for a German passenger aircraft, the VFW 614. This could reasonably be converted to a marine turbo-shaft engine, with an output of some 7000hp and an exceedingly good fuel consumption, though re-design of the low pressure compressor and turbine would be needed when the fan of the aero-engine was removed, Joint British-French-German discussions were held, and there seemed a good prospect that such an engine, with a broad European base, would have a ready market, both in the next generation of German and French patrol boats, and as a possible cruising engine in larger ships of all three navies. The French were also interested in using it for locomotive propulsion in projected high-speed trains. After a considerable period of debate, the project was finally abandoned, largely because the three

countries each had rather different requirements, and by this time the British had not only started to develop the Tyne, but were already looking ahead for a unit of considerably higher power — a search which was to culminate in the development of the Marine Spey.

A much more ingenious and radical approach was also proposed by Bristol Siddeley. This took the process of marinising aero-engines to its logical conclusion, basing design not merely on a single aero-progenitor, but building up a composite unit which would combine suitably matched components from several different sources. The proposed new engine, the Centaur, was originally conceived as using the high-pressure spool of the M.45. However, it avoided re-design and redevelopment of the low-pressure section by using an existing compressor from the Armstrong Siddeley Viper, which happened to be of the right capacity in terms of pressure-ratio and air-mass flow, and which had a great deal of aero-background and experi-

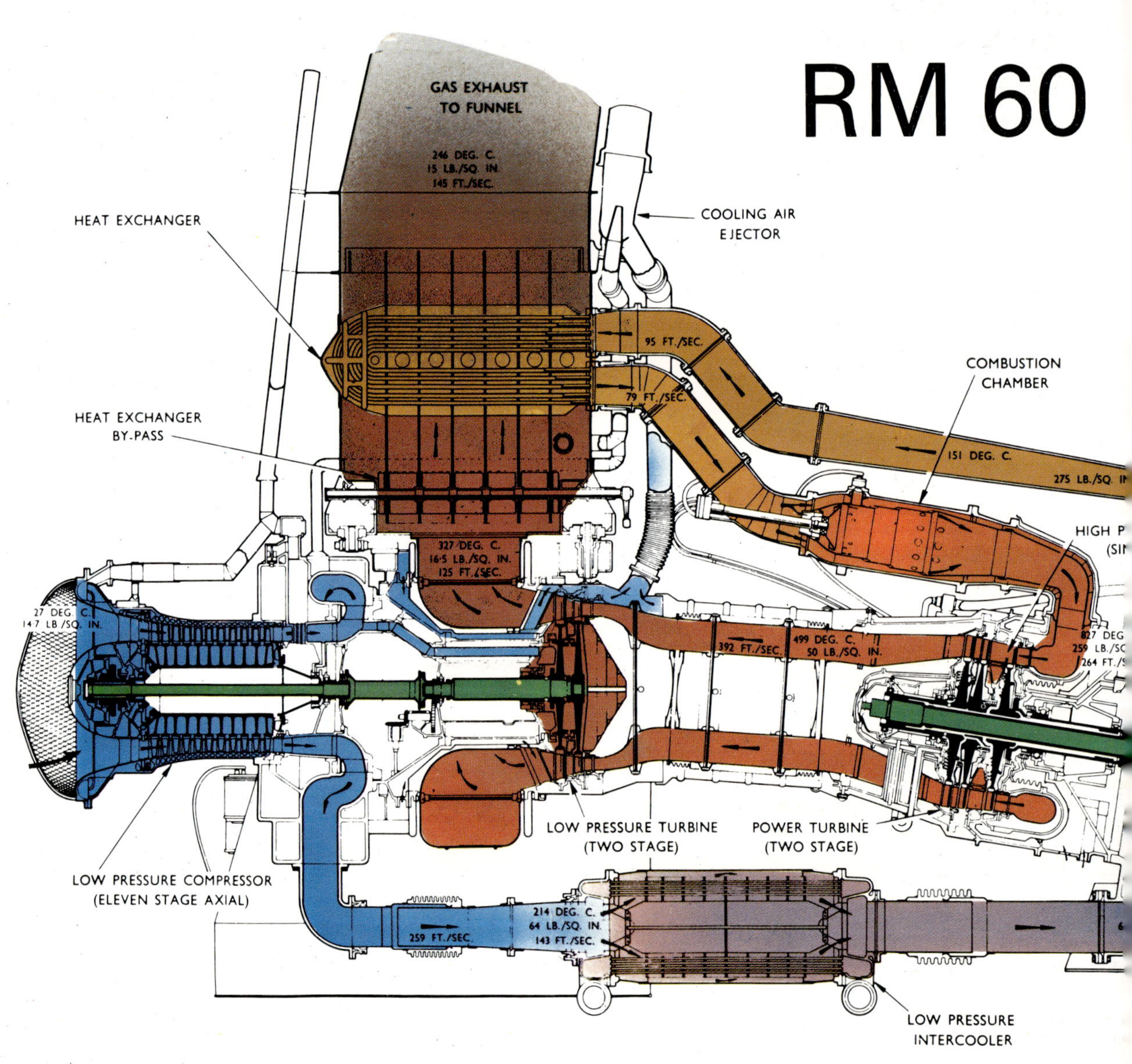

RM 60
GAS EXHAUST
TO FUNNEL
246 DEG. C.
15 LB./SQ. IN.
145 FT./SEC.
HEAT EXCHANGER
COOLING AIR
EJECTOR
95 FT./SEC.
79 FT./SEC.
COMBUSTION
CHAMBER
HEAT EXCHANGER
BY-PASS
151 DEG. C.
275 LB./SQ. IN
327 DEG. C.
16·5 LB./SQ. IN.
125 FT./SEC.
HIGH P
(SIN
27 DEG. C.
147 LB./SQ. IN.
392 FT./SEC.
499 DEG. C.
50 LB./SQ. IN.
527 DEG
259 LB./SC
264 FT./S
LOW PRESSURE TURBINE
(TWO STAGE)
POWER TURBINE
(TWO STAGE)
LOW PRESSURE COMPRESSOR
(ELEVEN STAGE AXIAL)
214 DEG. C.
64 LB./SQ. IN.
143 FT./SEC.
259 FT./SEC.
LOW PRESSURE
INTERCOOLER

The RM 60 used a highly complex cycle, with an unprecedentedly high compression ratio, and by the standards of those days it had a very good efficiency, with a specific fuel consumption 0.6 and 0.7 lb/hp-hr over nearly the whole power range.

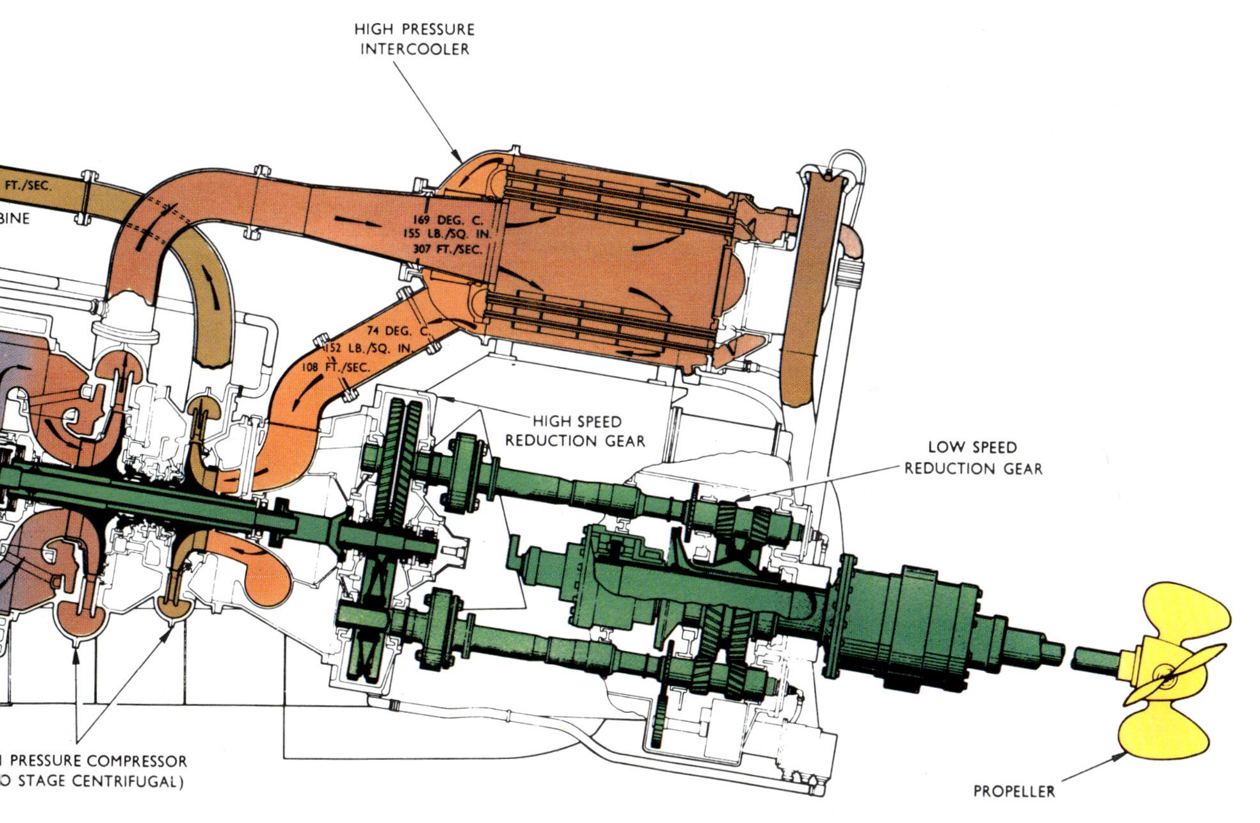

'BRAVE' CLASS

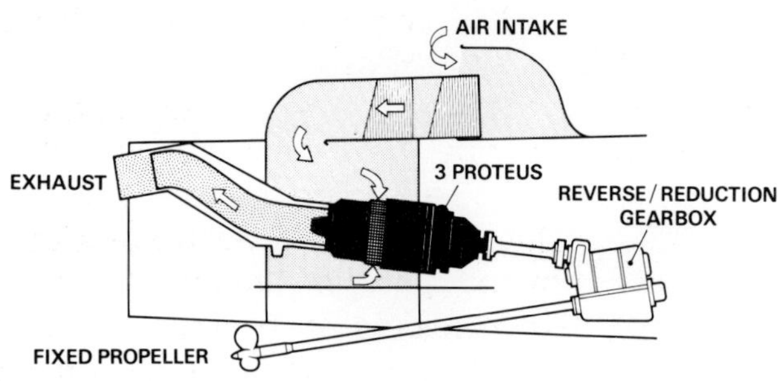

'SØRIDDEREN' CLASS

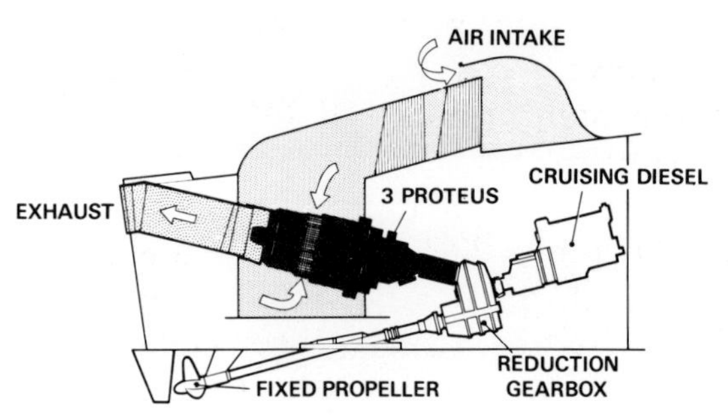

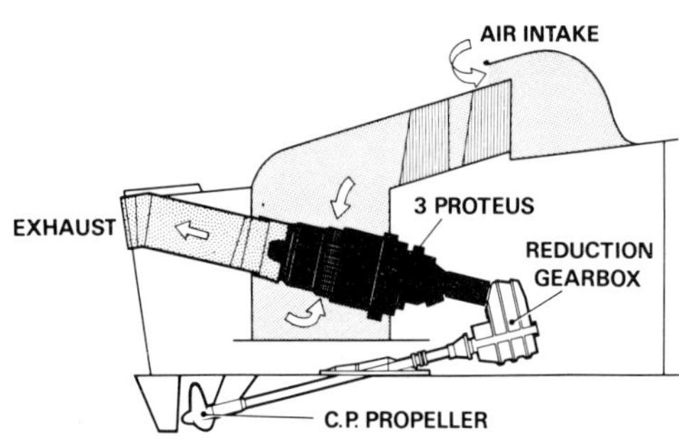

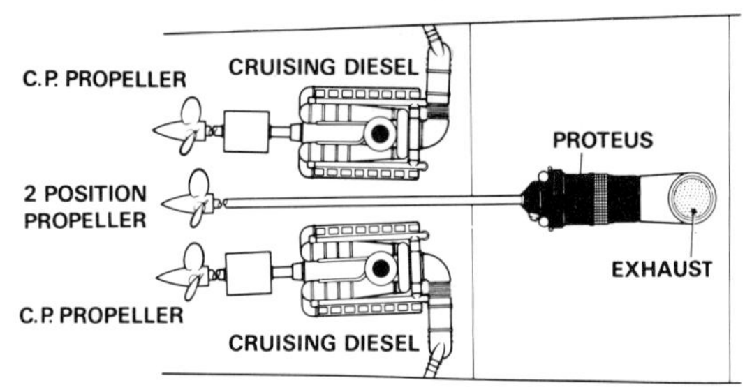

'SPICA' CLASS

'FRECCIA' CLASS

U.S.S. HIGHPOINT

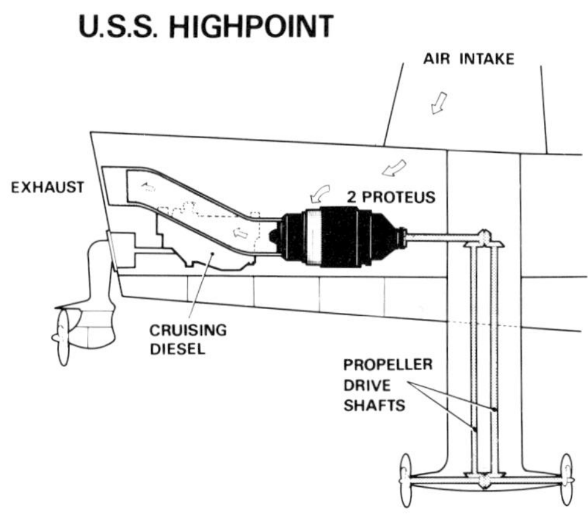
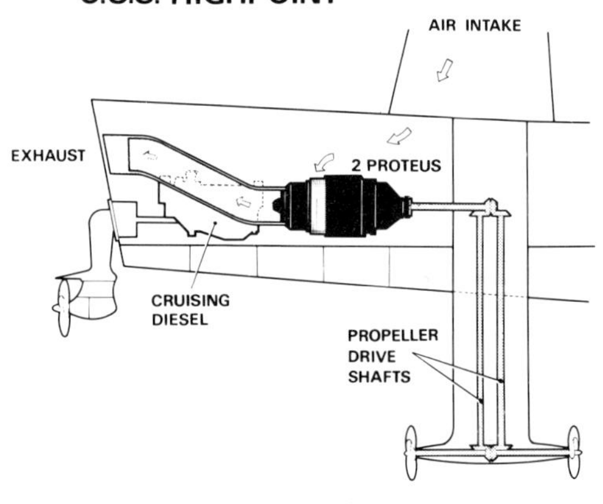

U.S.S. FLAGSTAFF

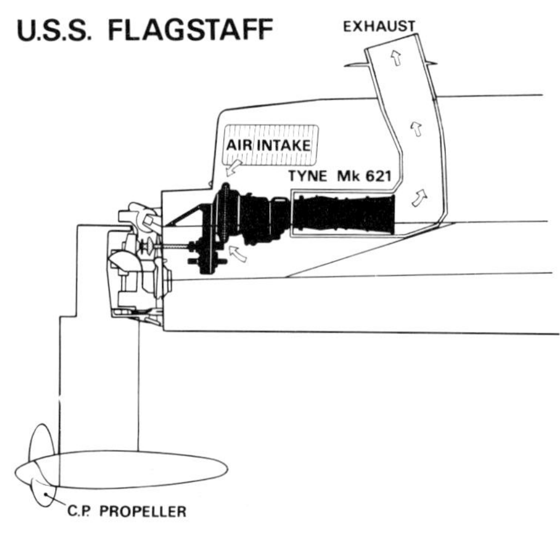

U.S.S. TUCUMCARI

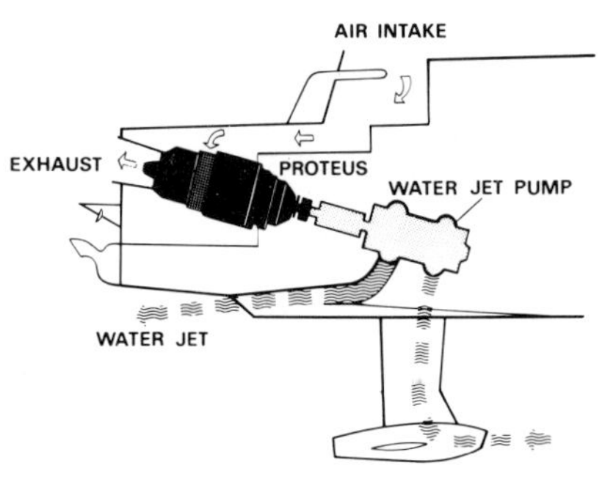

Proteus transmission systems. The V-drive, with short transom exhaust, produced a very compact installation. Simplest of all was the water-jet arrangement used in the Boeing hydrofoils, but this is less efficient than a propeller, except at very high speeds.

ence. Once the principle was established, it was possible to contemplate taking components from a large number of sources — suitably scaled down if necessary — and the Centaur design in a later form used the Viper 21 compressor, with the final stage removed, a scaled-down version of the high-pressure compressor of the Olympus 22R — the engine which powered the short-lived but still regretted TSR 2 strike aircraft — and high-pressure and low-pressure turbines derived from the Concorde engine, the Olympus 593.

This may sound a bit of a pig's breakfast but, relying on already-tested components, it should have been a relatively low-risk engine, requiring limited expenditure on development, and it promised a unit of 8000hp with a specific fuel consumption of 0.42lbs/hp/hr — a figure still not yet equalled for an engine of that size.

It was destined never to be developed. It could have been a very attractive engine for a larger generation of fast patrol boat, but there was no British requirement for such craft. It would have been ideal for larger hydrofoils and hovercraft; but in the 15 years which have elapsed since then no larger hovercraft have been built, and the only hydrofoils which might have used such an engine are the small *Pegasus* class of American PHMs.

Looking back from today, we can see that the Centaur, if it could have been developed successfully, would have been an excellent cruise-engine for the frigates and destroyers of the 1970s, but by the spring of 1966 the Ministry of Defence had embarked on testing of a marine version of the Tyne and was already considering this engine for cruise-propulsion.

So the Centaur went the way of all paper engines, but it did signal a valid line of approach to the problem of adapting aero gas turbines to particular marine requirements. The Proteus continued its career, and its adoption, during the 1970s, for no fewer than 28 new patrol boats, in Sweden, Denmark, and Yugoslavia, as well as for the Italian hydrofoils, belied the forecasts of its early demise.

By this time it was also well established in the larger hovercraft, but the engines for these merit a chapter of their own.

*U.S.S. **Tucumcari**, one of two prototype hydrofoils, built by Boeing for the United States Navy, with a single Proteus and water-jet propulsion.*

The Saunders Roe SRN-1, the first sea-going hovercraft. Very experimental, but highly successful. Propulsion was later provided by a Viper jet engine.

A HARD SCHOOL ~ THE HOVERCRAFT

So far we have been considering gas turbine applications in conventional patrol boats and hydrofoils. Compared to industrial applications, and the later use of the engines in big ships, the early small-craft installations posed severe problems, in operating in a confined space and in having to contend with a hostile marine atmosphere — a sufficiently difficult environment for their debut in the naval field — but the problems of patrol boats were mild compared with those which the hovercraft would bring.

At the same time that the 'Brave' class fast patrol boats were being planned, and development was starting on the marine Proteus, Christopher Cockerell was experimenting with his newly-invented hovercraft principle, and on 1st June, 1959, the world's first manned hovercraft, the Saunders-Roe SRN-1, made her maiden 'flight' in the Solent. For a first experimental craft she was extraordinarily successful, and she was soon flying freely on the waters of Spithead. Less than two months after the start of her sea trials she made history by crossing the English Channel, on 25th July, 1959 — the 50th anniversary of that first cross-channel flight by Bleriot.

At first, the SRN-1 was powered only by an Alvis Leonides piston engine, of 435hp, for she was built purely as a research craft to study hovercraft principles and problems, and the first concern was the supply of air for the cushion, which was provided by a four-bladed fan mounted vertically in the middle of the craft. For forward propulsion, air from the cushion plenum chamber was bled off into an aft-facing duct on each side of the craft, which 'blew' it forwards (these ducts were later re-designed, with valves at both ends, to provide reverse thrust for manoeuvring). By this means, one-third of the total air from the fan could be diverted from the cushion to forward propulsion, giving a speed of 25 knots. It was appreciated, even at the design stage, that there was a need to exper-

iment with higher speeds and to provide more power for propulsion. As often happens, initial plans had had to be tailored to a limited budget, but the SRN-1 was soon to be developed into a much more exciting vehicle, with a speed of over 60 knots.

For a craft designed for high-speed operation, with a high sensitivity to hull-loading and a need for a large amount of installed power, the lightweight gas turbine was an obvious choice. For best efficiency at speeds of around 60 knots, the most suitable form of propulsion was a gas turbine driving an aircraft-type propeller, and all the later Saunders-Roe craft used this form of propulsion. Some more recent American hovercraft have used directional air ducts for part of their propulsive power, and the Vosper hovercraft of the 1970s were driven by ducted propellors; but the first propulsion engines for the SRN-1 — again for reasons of economy — were aircraft jet engines. Before it had been on trial for a year, in May 1960, it was fitted with a small Blackburn-Turbomeca Marboré jet, giving 700lbs thrust, which raised the speed to 45 knots, and the following year this was replaced by the Bristol-Siddeley Viper, which provided 1500lbs of thrust and gave a maximum speed of 65 knots.

The Viper 5 used in SRN-1 was one of an illustrious family of jet engines which sprang from very modest beginnings, for the original requirement, for which the Viper was first produced by Armstrong-Siddeley in 1952, was for powering the Australian Jindivik unmanned target aircraft. This was expendable, and the specification called for a thrust of 1500lbs and a total engine life of 9 hours! Armstrong-Siddeley development engineers confessed that this was too difficult — it was impossible to design a reliable unit with a life of *only* 9 hours. Today, later marks of the same engine, which is still being produced for new applications, develop 4000lbs of thrust and run for 4000 hours between overhauls, and Vipers have been used in some 29 different types of civil and military aircraft.

For driving a hovercraft at 60 or 70 knots, however, a jet engine, with a high-velocity gas stream emerging from a relatively small tailpipe, is an exceedingly inefficient means of propulsion. What is needed at these speeds is a much larger but slower-moving column of air, and the successor to the SRN-1 was already being designed with aircraft propellers when SRN-1 trials were started in 1959. This new craft, the SRN-2, was no longer an obviously experimental vehicle, but intended rather for operational research — a much more finished design, and on a much larger scale, with the weight increased from 6 tons to 27 tons, air-pressure in the cushion from 25lbs/sq/ft to 75lbs/sq/ft, and a far greater power requirement. This was met by four Nimbus turbo-prop engines, each developing 700hp. These were being built for the Wasp and Scout helicopters by Blackburn Engines, by this time part of the newly-formed Bristol-Siddeley, which already had some experience of hovercraft operating-conditions through the running of the Viper in SRN-1, and could also draw on marine Proteus experience in conventional craft.

SRN-2 started trials in 1962, and in the same year another hovercraft, the 11-ton VA-III, was launched by Vickers, who had likewise been experimenting with the new system. A four-engined craft again, the VA-III was powered by Blackburn-Turboméca Turmo engines, developing 350hp, but with a different system of transmission. The VA-III used two engines for the lift fans, and two driving aero-propellers for propulsion; the SRN-2 introduced a combined transmission system, which has been used in all later hovercraft produced by Saunders-Roe, British Hovercraft Corporation* and Vospers. In this system, each engine (or each pair of engines in the case of SRN-2 and SRN-3) drives both a lift fan and an aero propeller, through bevel-gearing, and the sharing of power between lift and propulsion is determined by using a controllable-pitch propeller. Thus, by putting propellers into fine pitch, forward thrust is reduced, the loading on the propeller decreases, and the larger share of the engine power goes to increase fan-output. The controllable-pitch propeller also affords a simple method of manoeuvring by providing astern power in reverse pitch.

Both the SRN-2 and the VA-III carried out some limited passenger-carrying operations, but it was not until the advent of the 7-ton SRN-5 that regular passenger services became a practical possibility. Meanwhile, Saunders-Roe were engaged in designing a larger and more ambitious craft, which was to be used by the recently-formed Inter-Services Hovercraft Trials Unit (IHTU) for serious evaluation of a number of military roles. This was the SRN-3, of 35 tons, whicn was handed over to the IHTU in June 1964 and which continued to run for ten years, carrying out a wide variety of trials in naval and military applications, building up 1400 operating hours and covering some 40,000 miles over land and sea in German, Danish and British waters.

The SRN-3 used the same transmission system as the SRN-2, but increased power was required, and Bristol-Siddeley proposed using the Gnome engine. This was a unit of more advanced design, developed by De Havilland Engines (also by then part of Bristol-Siddeley) from the General Electric T-58 and used to power the Royal Navy's Sea King and Wessex helicopters. It was to become the standard gas turbine for all the smaller SRN-5 and SRN-6 hovercraft and over 120 units were to be produced in the years ahead.

The Gnome provided not only increased power but greater efficiency. The Nimbus, with an output of 700hp, had a specific fuel consumption of 0.84lbs/hp-hr, equivalent to an efficiency of 16 per cent. The Gnome went into hovercraft service with a rating of 1050hp and a

The British Hovercraft Corporation was formed in 1966 by a merger between Saunders-Roe and the Vickers Hovercraft interest.

fuel consumption of 0.625lbs/hp-hr, or 22 per cent efficiency. Over the years it has been improved, in the manner of gas turbines, and now delivers 1400hp with a consumption of 0.60lbs/hp-hr — a good figure for a small engine of this size.

The SRN-5 hovercraft, powered by a single Gnome, went into service in 1964 and was followed two years later by a stretched version, the SRN-6, with overall length increased from 39ft to 48ft, and weight from 7 to 9 tones, but still using the same power plant and still capable of 60 knots.

These small craft have seen an astonishing variety of service, civil and military, from the rivers of Vietnam to the Canadian Arctic, from the jungles of Borneo to the upper waters of the Amazon, from the navies of the Arabian Gulf to the Isle of Wight ferry service. They have done

oil surveys in Alaska; they have penetrated 200 miles into the Libyan desert — and they have subjected their Gnome engines to every sort of undesirable experience.

From the engine manufacturer's point of view, of course, this hovercraft experience, though it produced many headaches, has been invaluable. The hovercraft is unique in surrounding itself with dense clouds of spray even on the calmest day, and since it is fully amphibious, it is able to vary salt water with dry sand, or with a particularly nasty and difficult combination of the two. The effect of salt and sand on the engines was bound to be a serious problem, but it became doubly so when the much larger SRN-4 car-ferry hovercraft were introduced in 1968. Powered by four of our old friends, the Rolls-Royce Proteus gas turbines, they started to operate a regular, virtually non-stop cross-

The Vickers VA III. Powered by four Turmo engines, she was the world's first hovercraft to operate a passenger ferry service, between Rhyl and Hoylake, across the Dee estuary, in 1962.

channel ferry service, in which any serious delays in the schedule could not be tolerated.

Today, with six SRN-4 hovercraft in commission, operating a continuous cross-channel service throughout the year, and making, between them, fifty crossings a day during the summer months, the Proteus engines in these craft are accumulating over 30,000 hours running a year. They have built up a cumulative total of 370,000 hours and, more important from the engine operator's point of view, this type of service, with a half-hour journey-time, involves constant stopping and starting. However clever we may have become over the years, sand and salt cannot entirely be excluded from the engine. It says much for the progress that has been made in engine-design and in the development of new materials, as well as in air-filtration, that whereas engines were having to be removed for overhaul after 1000 hours in the earlier years, today they are allowed to run for 8000 hours between inspections.

Even in a big ship, the business of air-filtration is not simple. We have noted in an earlier chapter how the first marine gas turbines ran with virtually no air-intake filtration arrangements, and merely relied on the intake being sheltered by the bridge structure. This was clearly insufficient in bad weather conditions, and when the Brave class FPBs were designed the air-intake trunking incorporated two series of longitudinal splitters which, as well as acting as sound-absorbing panels to reduce the noise from the gas turbine air inlet, had a secondary function of collecting spray on their surfaces and draining it from their trailing edges into drainage "gutters" across the floor of the intake duct. This rather crude device worked remarkably well, by the standards of those days, and we were all greatly impressed. It had an efficiency of about 95 per cent. This meant that, under typical small craft conditions at sea, where one might expect a salt concentration of 10 ppm*, the amount of salt reaching the engine was 0.5ppm. This was accepted—partly because there was, at that time, no reliable means of actually measuring the amount of salt in the air at sea, or any reliable data to tell what the safe limit should be.

Salt deposition on the cold parts of an engine could become a problem, but seldom a serious one and usually it was easily cured. Much more intractable was the corrosive effect of salt on the hot turbine blades. The battle against this was fought — and still is fought — by four means: by keeping out the salt water, by choosing corrosion-resistant materials, by protecting blades with special coatings, and by limiting the temperature at which they operated. The established doctrine in the 1950s, and in the early days of the hovercraft, was that, always provided you designed your intakes carefully and kept temperatures below a safe level (which was never entirely defined or agreed) you had no problem. Initial running of the Proteus for 225 hours on the test-bed in 1955, with salt water sprayed into the intake at 1ppm, had had no noticeable ill-effect.

As soon as experience began to build up, however, it was realised that the game was not quite as easy as had been anticipated. Under bad weather conditions, heavy spray could greatly increase the amount of salt ingested; there was pressure to increase temperatures in order to get more power, and in any case temperatures, in service, showed a considerably 'scatter' above and below the theoretical figure; protective coatings were not entirely reliable, and blade materials which had the greatest high-temperature strength were often those most susceptible to corrosion.

After a long and painstaking period of experimentation and development, which was shared by Rolls-Royce and the National Gas Turbine Establishment and tested at sea, it was concluded that a safe limit for salt was 0.01ppm; that this required a filter efficiency, not of 95 per cent, but of 99.9 per cent; and that this could be achieved, at least in large vessels, by proper siting of intakes and by using the type of three-stage filter system which we have in service today.

This by no means solved the problem for hovercraft, however. In SRN-1 the salt entering the air intake in the original arrangement was nearly *1000ppm* when the craft was hovering,

ppm = parts of dry salt per million parts of air, by weight. For comparison, the amount of salt in a normal marine atmosphere, without allowing for any locally generated spray, is only 0.26ppm in a 20-knot wind, but with a 40-knot wind it rises to 50ppm.

"SRN 3, 35 tons and powered by four Gnome engines, was the principal craft operated by the Interservices Hovercraft Trials Unit for military evaluation."

The SRN 6, ''stretched'' version of the SRN 5, which has operated in many parts of the world, and is in naval service today in Egypt, Saudi Arabia, Iraq, and Iran.

"A tough testing ground for both engines and intake filters ...."

".... but sand could be an even nastier problem."

even though it dropped to a mere 20ppm at 20 knots forward speed, when the spray curtain was largely left behind. SRN-2 was not much better, and it was, of course, impossible to site hovercraft intakes in the shelter of the bridge, because there was no bridge, and, when the craft was hovering, no shelter.

The most promising development was the use of Knitmesh panels — a slightly misleading proprietary name, since they were not knitted and had no regular mesh, but consisted of a coarse mat or pad of Monel metal or polypropylene 'wool', rather in the manner of a saucepan scourer. These panels, which form an integral part of virtually all marine air-filtration systems today, are extremely efficient in coalescing small particles of water into larger droplets which can then be carried away through normal drains. They were less effective in dealing with heavy concentrations of water, but it had been found that the more solid spray could be removed relatively easily from the incoming air by using centrifugal or inertial separation methods, passing the air at high velocity through a series of curved vanes, so that the heavy water droplets were thrown to the outside of the curve and could there be collected and drained. Such is the present system — a combination of inertial and coalescer filters — but Rolls-Royce and British Hovercraft Corporation found that they could improve on this in one respect, which was to prove of great importance in the future development of hovercraft engine installations.

In 1965, the IHTU took an SRN-5 hovercraft to Aden to carry out trials over desert sand. It was realised that the knitmesh filters, so excellent for water-separation, would be useless in the desert, since dry sand merely passed straight through them; so a complex inertial filter was fitted, in the form of a bank of Strata tubes. These were small plastic tubes with internal spiral vanes which imparted a swirl to the incoming air and threw the sand particles out to the periphery of the tube, where they could be collected and separated from the main air flow. This was a useful system. It had an efficiency with solid particles about equal to that of the earlier 'Brave' class splitters with water particles — some 94 per cent of sand was filtered out of the air stream.

A hovercraft, however, creates a cloud of sand considerably denser than the average desert sandstorm, and particles of sand travelling at high velocity — the Gnome compressor runs at 25,000rpm, with a compressor blade-tip speed of 640 miles an hour — will demolish small steel compressor blades very effectively. The first two engines lasted six hours each.

The hovercraft itself furnished a partial solution to its own problem, since it had an enormous centrifuge already to hand in the shape of the lift fan. It was found that the larger part of the sand ingested by the fan was thrown towards the base of the fan casing, and by taking air for the engine from the relatively clean area at the top of the plenum chamber the designers could, as it were, get rid of a large part of the problem before they started. By following this initial centrifuge with a double bank of Strata tubes, as well as a Knitmesh filter stage, it was possible, with dry sand, to increase filtration efficiency to over 99 per cent.

Two years after the ill-starred trip to Aden, the IHTU took the SRN-5 to Libya. In the course of this tour, the craft operated for 32 hours over sand and rock, penetrating nearly 200 miles into the Libyan desert to the Great Sand Sea south of Gairabub, with a large part of the journey carried out at 40 knots over appalling terrain. These conditions, with sharp rock as well as sand, and frequent steep-sided wadis to cross, can play havoc with a hovercraft's skirt; but with some patching *en route,* the SRN-5 completed the return journey successfully — a tribute to her designers and her crew.

The engine was inspected after returning to England and was found to be in excellent condition, good for another 300 hours running.

When the big SRN-4 hovercraft were introduced into the cross-channel service in the following year, they were fitted initially with a conventional Knitmesh intake; but the lessons learnt in the smaller craft could now be applied. This still involved a painstaking process of trial and modification, but it was to culminate in an intake arrangement which was capable of coping with constant operation over water and a tidal shore-line, and, in combination with further improvements in the Proteus, of providing an engine in which even the hot turbine section could run for over 3000 hours without inspection.

This prolonged passenger service, although

ABOVE: *SRN 4, the familiar cross-channel ferry.*

TOP RIGHT: *As in all the standard BHC hovercraft, SRN 4's engines each drive both a lift fan and a propeller. The four-engine arrangement, with swivelling and reversing propellers, gives exceptional manoeuvrability.*

RIGHT: *In VT 2, each Proteus drives twin lift fans and a shrouded C.P. propeller.*

The SRN 5. Powered by a single Gnome engine and — with its "stretched" version, the SRN 6 — the standard small hovercraft in the BHC range.

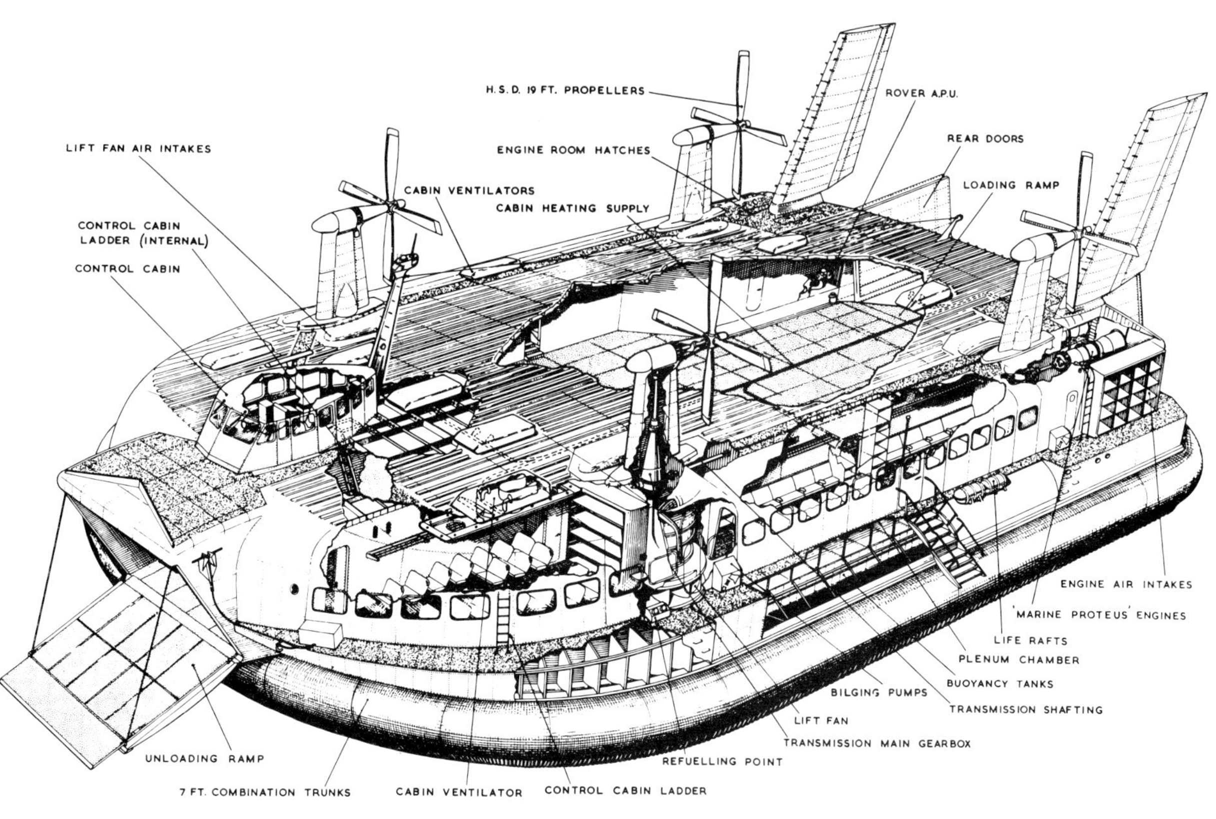

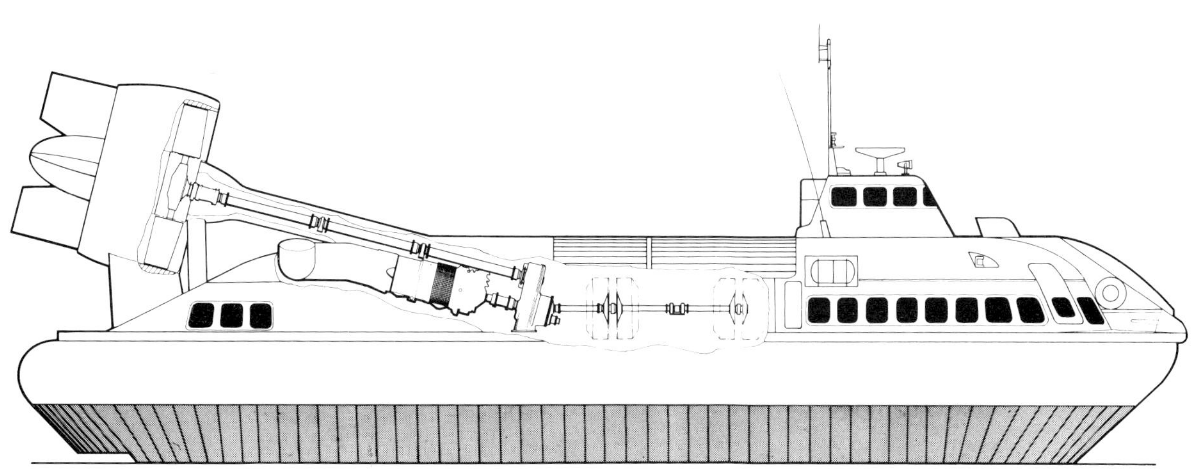

VT2 HOVERCRAFT MACHINERY INSTALLATION

it may have lacked the operational variety and the more extreme conditions associated with naval or military applications, did mean that sea-experience was being accumulated at a tremendous rate. Not only could improved intake arrangements be quickly tested at sea, but modifications to the engines themselves, and new turbine blade materials, could build up running time in operational conditions at a rate of 1500 - 2000 hours a year, more than three times as fast as in the average naval craft of this size.

Hovercraft of the size of the SRN-4 — the latest version 185ft long and weighing 300 tons — have not been used in a military role, although there is an obvious application as a landing craft, if the need were to arise. Two intermediate craft were, however, to be developed more specifically for military evaluation as fast attack craft, amphibious assault craft, for coastal interception, for logistic support or for mine countermeasures. These craft, both powered again by the Proteus, were the single-engined British Hovercraft Corporation's BH-7, of 55 tons, the first of which joined the IHTU in 1970, and which is also in service in the Iranian Navy, and the Vosper-Thornycroft VT-2,

of 100 tons, with twin Proteus driving a new type of shrouded propeller which, as well as being less vulnerable to damage, is considerably less noisy than the normal aircraft-type propeller.

So we now have a series of well-developed hovercraft, of 10, 55, 100, 200 and 300 tons, which could be used for military purposes, and which have been extensively evaluated. To some extent they have suffered the same fate as the earlier fast patrol boats, in that they have been successful craft for which the Royal Navy has as yet no specific application. It is difficult to see an ASW role for them, but as attack craft they have the advantage of unrivalled high speed. They do have obvious virtue as minesweepers which are largely isolated from the effects of underwater explosions — if appropriate high-speed sweeping gear is available — and as amphibious assault craft which can carry on inland, beyond the line of normal shore defences. The Trials Unit, taken over by the Navy in 1974, and now NHTU, continues its evaluation of the BH-7 and the large VT-2, both powered by the now middle-aged but still unrivalled Proteus gas turbine.

VT 2, the major Vosper hovercraft, undergoing trials with the N.H.T.U. Of 100 tons, and powered by two Proteus engines, she has considerable military potential.

THE BIG SHIP DEBATE

We have already noted that, in 1946, the British Admiralty, in addition to starting development of lightweight gas turbines for small craft, commissioned an engine of a different stamp, the English Electric EL 60A, as a prototype for big-ship propulsion.

This was an extremely imaginative decision, for, while the prospect of gas turbines in small craft was opening up new vistas and new possibilities, their use in larger vessels was at best an interesting experiment, which appeared to confer no valuable advantage.

In small craft, the comparison was with diesel engines, and it could be shown that for limited range the gas turbine had a clear advantage in machinery weight and space, in total machinery-plus-fuel weight, in ease of operation, reduction in maintenance, saving in onboard manpower, high-speed performance, and instant readiness. For hovercraft and for high-speed planing craft, in both of which hull loading was a critical factor, the gas turbine's light weight alone was sufficient to make it an obvious choice, even in the earlier days when engines available were still relatively inefficient and far from reliable.

When the debate moved to machinery for frigates or destroyers, many of the gas turbine's advantages disappeared and the remainder became of much less significance. Above all, there was no need to go to gas turbines, since steam plant was already providing an entirely adequate means of propulsion and there was ample evidence that it could still be improved, in efficiency, flexibility, and reliability, as well as in ease of maintenance and automation, to meet any future needs in the fleet that could be foreseen.

In the matter of machinery weight, there is, of course, an enormous difference between the requirements of small and large craft, because of the large variation in specific power with displacement. Even assuming that both types of craft are operating at the same speed, the specific power, in horsepower-per-ton, decreases rapidly as displacement increases, so that whereas a 100-ton fast patrol boat will require 5000hp to drive it at 30 knots, or 50hp/ton, a ship of 3000 tons will require little more than 30,000hp for the same speed, or 10-12hp/ton. Since the small craft are in practice required to have a top speed of 40 knots or more, and the larger are content with 30 or less, then — whether or not there is any real tactical case for this speed distinction — the patrol boat needs 100 horsepower per ton and the frigate only one-tenth of this figure. In consequence, while lightweight machinery could make a big difference in the patrol boat, it was of much less importance in major war vessels.

The class of gas turbine being considered as a substitute for steam plant — as a main propulsion engine — was typified by the EL 60A. It was felt necessary to use a complex cycle, with large and heavy heat exchangers, in order to achieve reasonable efficiency both for cruise and for high speed, and it was obviously necessary to have a robust, heavily-built unit. The result was that there was little or no saving in weight. In parenthesis, it may be noted that this was not necessarily counted as a disadvantage in some circles. A little later, the idea of using lightweight engines aroused real misgivings amongst some naval architects on account of its effect on ship-stability, and low machinery-weight is still cited as one of the 'problems' associated with gas turbines. This is in part because reduction in machinery weight has been accompanied by an increase in top-weight, through the introduction of ever larger and heavier radar antennae, and some loss of weight low down, in shell rooms, through the replacement of heavy gun armament by missiles. It was even argued by some protagonists in the debate, as a final clincher to demolish the gas turbine case, that it might be necessary to resort to the expedient of installing fixed ballast to maintain stability — to which the correct answer, of course, is: 'So what?' Fixed ballast is cheap, unobtrusive and easily installed, and has been used in ships since the 19th Century.*

Space in a warship is always at a premium, and here the gas turbine might be expected to have an advantage; but although designers would ultimately be able to do away with boiler rooms, they were a long way from abandoning steam altogether in the early days. Some boiler-capacity would still be required, and, in any case, an engine such as the EL 60A occupied almost as much space as equivalent steam plant, including boilers. Moreover, while a gas turbine installation might save space down below, it had one disadvantage compared to steam plant, in that it relied on the intake air not only for combustion of the fuel, but also for cooling the resultant hot gases to an acceptable level for the turbine blades, and in consequence required a large volume of air. So the uptakes and downtakes would occupy much more space than those for an equivalent boiler, and would partly offset the saving in machinery space.

Thus saving in weight and space was not quite such a strong point in the gas turbine's favour as had at first appeared, particularly if heavy gas turbines with complex cycles were to be used. Nevertheless, if a simple gas turbine could be used, as a boost engine in combination with steam machinery, there was some real gain to be made in the matter of weight and space. The use of gas turbine boost engines had been suggested in earlier days — for instance in the proposals for combined steam and gas turbine machinery (already mentioned in Chapter 3) which were outlined by the Americans in

As late as 1981 it has been claimed in a TV discussion that RN frigates are unstable, because of gas turbines.

1950 — and the saving in weight and space was, in fact, one of the main arguments put forward when the Admiralty finally decided to adopt gas turbine boost engines.

As far as fuel was concerned, it was assumed, in those early days, that steam plant would continue to be more efficient than the gas turbine, and the latter suffered from the further damning disadvantage that it required distillate fuel, and could not tolerate the heavy residual fuel then used universally for boiler-firing. It was widely agreed that, if the gas turbine was to find a permanent place in warship propulsion, one of the first problems was to adapt it to furnace fuel. It is, in fact, a problem which has never been solved, for today's marine gas turbines are less able to tolerate heavy residual fuel than their relatively crude and low-temperature predecessors. As far as naval machinery was concerned, however, the problem disappeared in the 1960s, when the Royal Navy, and most other navies, decided, for reasons quite unconnected with gas turbines, to switch to distillate fuel for boiler-firing, in order to reap the big advantage of improved reliability and reduced boiler maintenance. So the heavy fuel lobby retired from the naval battle, though they have continued to fight on the industrial front.

The gas turbine promised some reduction in day-to-day maintenance, as it had less auxiliary machinery than steam plant, but again this was not a particularly strong point, since it was always assumed that, while main steam plant would last the life of the ship, any gas turbine would require to be opened up and repaired at relatively short intervals, and significant repair-time would also be required for heat-exchangers. Savings in day-to-day maintenance would therefore be offset by periodic overhaul and corresponding reduced availability of the ship.

One clear advantage the gas turbine did have: it provided quickly available power, without the need to keep steam up, and a ship with at least a proportion of gas turbine power could therefore get under way almost immediately, even when it was at extended notice for steam. This is particularly the case with the lightweight aero-derived engine, but even the heaviest and most complex units can acquit themselves well. The EL 60A was designed to be able to reach full power from cold in 20 minutes.

It was this ability to provide power at short

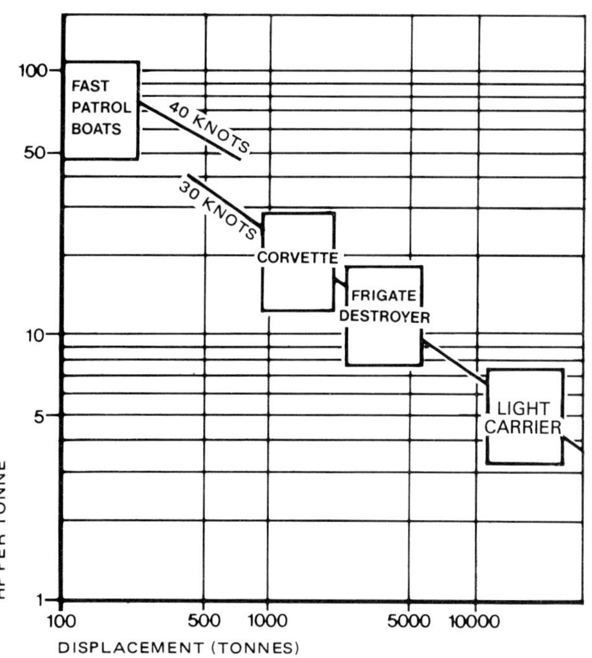

Specific power versus ship displacement. The horsepower-per-ton required to drive a ship at a given speed decreases rapidly as the size of the ship increases. In small craft machinery weight and space may be crucial, but in larger ships they become less important.

notice that was perhaps the most important advantage of the gas turbine boost engine of the 1960s, even though it was not at first appreciated, and was still regarded as something of an incidental bonus until its value was demonstrated in service.

However limited the advantages, and however inconclusive the arguments, in 1953 the Admiralty put in train a detailed investigation, by the Yarrow-Admiralty Research Department (Y-ARD) and Metropolitan-Vickers, into gas turbine boost plant for major warship propulsion. As a result, it was decided to use gas turbines, in combination with steam cruising engines, in two new classes of ship, the 'Tribal' class (Type 81) general-purpose frigates and the 'County' class guided missile destroyers (DLGs).

After much debate, it was decided that the frigate machinery should consist of 12,500hp steam, for normal cruising and manoeuvring, and 7500hp gas turbine power, on a single shaft. It will be noted that this division of power provides very limited 'boost': assuming a top speed of 25 knots and a normal power-speed relationship, the steam plant would give a maximum of 21-22 knots and the gas turbines would only increase this by 3-4 knots. But al-

though the gas turbines were originally intended primarily to give limited boost-power, it was appreciated that their quick-start capability would also make them valuable for manoeuvring, and reversing was therefore provided in the main gear box for the gas turbine drive, though the normal means of manoeuvring was to be by an astern steam turbine. Design of the engine, the G.6, was started by Metropolitan-Vickers in 1954, and by early 1958 a complete ship installation, including engine, gear box, and controls, was being run in a new test-plant built for the purpose at the Company's works in Manchester.

For the guided missile destroyers a two shaft arrangement was required, with 15,000hp gas turbine and 15,000hp steam on each shaft. Metropolitan-Vickers originally proposed for this a single engine, known as the G.5, but after studying the operational requirements, and the machinery space layout, it was decided to use two units of 7500hp, which also meant that the same unit could be used for both the frigates and the destroyers. The transmission arrangement was similar to that in the general-purpose frigates, allowing independent operation and manoeuvring on either steam or gas turbine.

In November, 1961, HMS *Ashanti,* first of the 'Tribal' class, was handed over to the Navy. She was followed a year later by HMS *Devonshire,* the first of the 'County' class, and during the next ten years a total of eight 'Counties' and seven 'Tribals' were built. There were some doubts about the reception that the new gas turbines would have at sea, but, from the first, they proved extremely successful and extremely popular, for they not only provided instant, remote-controlled power, but were very reliable and, above all, they demonstrated that gas turbines required extremely little onboard maintenance compared with steam plant.

During the 1960s, G.6 gas turbines were also installed in three Italian ships, the reconstructed destroyer *San Giorgio,* and two new frigates, the *Alpino* and *Carabiniere.* These ships mounted one G.6 on each shaft, but they dispensed altogether with steam, using diesel cruise propulsion in a CODAG* arrangement.

This was a novel departure for a ship of this size and power. It had been used once, a little

See Glossary, p. 112.

earlier, in Germany, the home of the diesel engine, for the *Köln* class frigates, of which the first commissioned in 1961, almost at the same time as *Ashanti*. The Italian ships used the two G.6s, totalling 15,000hp, in conjunction with four diesels giving 16,800hp; but the *Köln* class increased the gas turbine proportion of the total power, using two Brown Boveri units, each of 13,000hp, with four 3000hp MAN diesels. This was roughly in line with the sort of cruise-boost proportion that had been envisaged in the early American study, which had predicated 9,000hp steam and 21,000hp gas turbine; but even with these combinations the gas turbine boost engines would, in theory, only be used for a tiny fraction of the total running time. In practice, the British ships tended to use the gas turbines for a larger proportion of the time, simply because they were so readily available and so easy to operate; but on economic grounds there was a good case for providing a much greater proportion of the total installed power from the boost engines, and it was becoming apparent that there would be a need for a larger unit, in the 20,000-30,000hp range, if full advantage were to be taken of the new means of propulsion.

The G.6 was essentially a heavy engine, even though it had come from a line of earlier units which were basically aero-derived. The reason underlying the design was that, since the engine was to be a vital component in two important new classes of major warship, it was essential that it should be as nearly right as possible from the beginning, for it was clearly out of the question, with such a relatively small number of engines, to finance the kind of elaborate long-term development programme which was already recognised as being necessary for aircraft engines. This meant using relatively conservative design criteria. In addition, the engine had to be robust enough to withstand high shock-loads from underwater explosions, and it was generally accepted that plain journal bearings and conventional thrust bearings should be used. Much heavier compressor blading was used than in the earlier lightweight engines, in order to avoid, by brute force as it were, the sort of failure due to compressor-stall which had occurred in the G.2 gas turbine.

The result of this design philosophy was an engine which did prove, from the outset, to be extremely reliable, but which weighed 19 tons. In addition, conservative design in terms of pressure-ratio and temperature resulted in a relatively low efficiency, but since the engine was for boost, and intended only for very limited running in high-speed operation or manoeuvring, this was of little consequence.

The G.6 only suffered one serious failure, which occurred during the later stages of the initial trials in *Ashanti*. This was caused by an insufficient cooling-air supply to the first-stage turbine disc, and the fault was corrected without difficulty. It did, however, point an important lesson, for *Ashanti* was out of action for a long period while the engine was opened up and inspected, modifications made, and repairs, effected.

It was noted that small craft such as the 'Brave' class would not need to suffer the same delays, since their Proteus engines could be removed, and replaced by spare units, in a matter of hours. By this time, the 'Braves' had been in service for four years, and the Navy was beginning to revise its views on the robustness of aero-derived gas turbines. In particular, it appeared that ball- and roller-bearings were less vulnerable to damage than had been expected, and it was coming to be appreciated that the 'fragile' aero-engine, in its native habitat, did have to put up with a good deal of rough treatment.

Bristol-Siddeley, offering a marine version of the Olympus engine, were already suggesting the sort of compromise which has since become standard practice in all the larger British gas turbines — using an aero-jet engine, suitably modified, but mating it with a purpose-built power turbine which could be built as a really rugged unit, designed to last the life of the ship and largely immune to damage. The 'Olympus', of course, would also have the advantage of furnishing a unit in the 20,000-30,000hp range which, as we have noted, was already foreseen as about the right size for new boost-engine requirements.

Before a decision could be made, a number of questions had still to be answered. The lightweight aero-derived gas turbine had some obvious advantages and some obvious disadvantages as a prime mover for the specific job of warship propulsion. But it is over-simplifying merely to regard it as an alternative to the heavy 'steam-derived' gas turbine, and to compare one type with the other, since there was no question of *having* to choose either. Steam plant was already meeting all the Navy's needs and could certainly continue to do so; two classes of all-diesel frigates, the *Leopard* and *Salisbury* classes, were already in service; a new diesel, in which the Admiralty had great hopes, was under development by Rustons. The decision to be taken, therefore, was, first, whether there should be a change from the still current policy of relying largely on steam, and second which of several alternatives that change should embrace.

The decision which was finally taken was the culmination of a series of steps, some logical and some intuitive, and not all the arguments in the case could be foreseen in the early stages.

The whole process, however, did constitute a considerable revolution, so it is worthwhile examining in more detail, and with hindsight, we can condense a series of arguments which took place over a period of some years.

What are the requirements for warship machinery? First of all, if a ship is to carry the maximum weapon load for the least cost, steam (for want of a better word!) at 30 knots when required, but spend most of its time at half that speed, we can lay down a few basic rules for the machinery. The bigger the hull, the higher the cost; the greater the machinery load, the less the weapon load; so the first need is for minimum weight and space. There is also a need for high power, by normal merchant ship

*H.M.S. **Zulu**, one of the 'Tribal' class frigates, powered by one G.6 gas turbine of 7500 hp and a 12,000 hp steam turbine.*

standards, but for plant which, for reasons of fuel economy, will spend most of its time operating at much reduced power. In practice, this results in ships cruising at about half maximum speed, which means about one-eighth of full power. (It is assumed that the ship-designer is faced with an intransigent Naval Staff who insist, like most Naval Staffs, on having a speed of 30 knots, but who accept that the laws of hydrodynamics will make life increas-ingly difficult for them if they want to go any fas-ter).

Beyond these basic rules, there are a number of requirements peculiar to the warship's role, plus a few which are obviously desirable in any vehicle. We might summarise these under four main headings.

First, our machinery must be able to stand up to war conditions. This implies rugged con-struction, but, in particular, a minimum of vul-

*H.M.S. **Devonshire**. The 'County' class "super destroyers", of 6500 tons, were the first RN ships to carry missile systems (Seaslug and Seacat SAM). Powered by two G.6 gas turbines and a 15,000 hp steam turbine on each shaft.*

nerability to the specific hazards of a warship's life:

- Action damage
- Shock from underwater explosions
- Blast from above-water explosions
- Salt corrosion
- Nuclear fall-out

Second, it must have an unusual degree of flexibility, with fast starting, rapid manoeuvr-ing, high maximum speed, the best possible en-durance at highest possible cruising speed, and perhaps, also — a quality peculiar to warships — the ability to operate continuously at very low speed.

Third, propulsion machinery must have high availability. This might seem self-evident for any piece of machinery, but warship propulsion is a special case. True, if a merchant ship breaks down, you may lose a lot of money; but if a war-

The **Alpino**, one of two Italian frigates commissioned in 1968, with CODAG machinery comprising two G.6s with Fiat diesels.

ship breaks down, you may lose a war; and while a warship with one gun or missile system out of action is less effective, one with no main engines is very severely limited. So we need good reliability and the ability to continue in service for long periods, plus a minimum need for onboard maintenance and minimum time out of action in the event of damage — in other words, machinery that can quickly be either repaired or replaced.

In addition to these, there are a few external requirements which affect ship-operation:

> A low level of underwater noise (for anti-submarine operation)
> A low level of airborne noise
> A smoke-free funnel — and, in addition,
> A low rate of infra-red emission

When we come to examine the various machinery alternatives, the first thing we note is a strong polarization of pros and cons between the aero-derived gas turbine at one end of the scale and conventional steam plant at the other. On a number of counts, there is no doubt that the lightweight gas turbine is either outstandingly good, or at least demonstrably better than anything else:

1. It occupies least space and weight
2. It has the highest power-weight ratio
3. It is the quickest-starting and most rapidly manoeuvrable
4. It will run happily (though uneconomically) at very low speed
5. It has to have been developed to a very high degree of reliability for aircraft use, and we should be able to take advantage of this, even though we are changing the application and the environment
6. It requires almost no day-to-day maintenance
7. If it is damaged, it can be replaced in a few hours. Major repairs are done under workshop conditions, and not in the confined space of a ship's engine room
8. It also has the advantage over the 'heavy' purpose-built gas turbine, that the greater part of the very large development cost for a new engine has already been covered, even though some money will still need to be spent on necessary re-design and specific 'marinisation'.

Running time in service between overhauls had been considered one of the inherent problems in the early days, but by 1963, when machinery options were being reviewed, experience with the 'Brave' class fast patrol boats, and the great strides made in extending aero-engine 'life', gave promise that future lightweight gas turbines would prove quite adequate in this respect.

So far, so good. But the simple-cycle gas turbine is emphatically not economical when cruising at low power, and when we come to consider ability to stand up to war conditions, the aero-derived engine would seem, at first sight, to be an extremely unpromising candidate. It certainly is not of rugged construction if by 'rugged' we mean massive or heavy; but it has, of course, to have sufficient structural strength to withstand the stresses of its quite violent internal functions, as well as a considerable degree of buffeting, and accelerations up to some 10G, in normal aircraft use. Some of its external dressing may seem flimsy, but short of being actually kicked by irate sailors (which one can guard against by housing it in a separate compartment), these are unlkely to be affected by normal ship conditions if they have been designed for aircraft use, and some components will, in any case, require modification to suit them to a changed role. The gas turbine's external casings are light, but light weight does give the whole engine the advantage of accommodating more readily to abrupt changes in temperature caused by rapid increases and decreases in power, and of thus avoiding the thermal shock problems associated with more massive structures.

Even from the point of view of physical shock, 'lightweight' means that mounting structures can be lighter and simpler.

Certainly, if casings can withstand self-generated internal pressures, and the variations produced by aircraft flight conditions, they can also withstand the additional pressures imposed by external blast, while as far as action damge is concerned, the gas turbine does have the advantage, compared with steam machinery, of presenting a far smaller physical target and eliminating the large vulnerable area of boilers and steam lines. This is one advantage which had been attributed to the steam turbine, 60 years before, when it, in its turn, was being

METROPOLITAN-VICKERS GAS TURBINES

Designation	G.1.	G.2.	G.2/II	G.4.	G.6.
Year of test	1946	1951	1955	1956	1958
Max. power, s.h.p.	2500	4800	4800	5000	7500
Specific fuel consumption, lbs/hp/hr.	1.06	0.82	0.82	0.68	0.77
Compression ratio	3.5	4.0	4.0	6.3	6.3
Turbine inlet temp., degrees K.	1022	1072	1072	1089	1066
Engine weight, lbs.	4030	6950	6950	8160	41,440

Progress in gas turbine development: the range of Metropolitan-Vickers engines. The G.3 and G.5, not included above, were design studies only. The G.3 was a 12,000 hp lightweight engine using the 'Sapphire' as a gas generator; the G.5 was the 15,000 hp marine engine orginally considered for the 'County' class cruisers, which eventually used twin G.6s.

compared with reciprocating machinery.

In this category of physical robustness, the two things which caused most concern to marine engineers in the early 1960s were ball-bearings and low shock-resistance. The other weak points of the aero-gas turbine — its limited 'life' between overhauls and its susceptibility to salt corrosion — were by now being discounted, as 'lives' quite adequate for a boost-engine application were already being demonstrated, while the problem of salt corrosion, which was being tackled with some success even in hovercraft, would certainly not be so acute in a big ship.

In 1963, a detailed study of the Bristol-Siddeley Olympus was started by what was shortly to become the Ministry of Defence (Navy) — the world's last remaining Admiralty was relegated to the history books in 1964. There was as yet no commitment to the use of the Olympus. An examination was needed first, to determine what changes would be required to make it a fully 'marinised' engine, whether these were feasible, and, in particular, whether bearings could be modified to meet naval requirements and to what extent shock-resistance could be increased. From this study it was concluded that, with two main thrust bearings of increased capacity, the bearing system could be relied upon to have an adequate life in naval service under sea-level operating conditions. It was also concluded that, by strengthening engine casings, particularly in the region of the main bearings, and by the use of suitable mounting arrangements, the lightweight gas generator could be adapted to meet the full naval shock-requirements. It was agreed that a power turbine of heavy conventional design could be produced, which would stand up to all the stresses to which it might be subjected, either from external shock or from ship's propellers and transmission system.

In 1964, a prototype Marine Olympus gas turbine was ordered by the Ministry of Defence.

The gas generator of the Marine Olympus. This was developed from the Olympus 201 jet engine, which powered the Vulcan Mark II V-bomber.

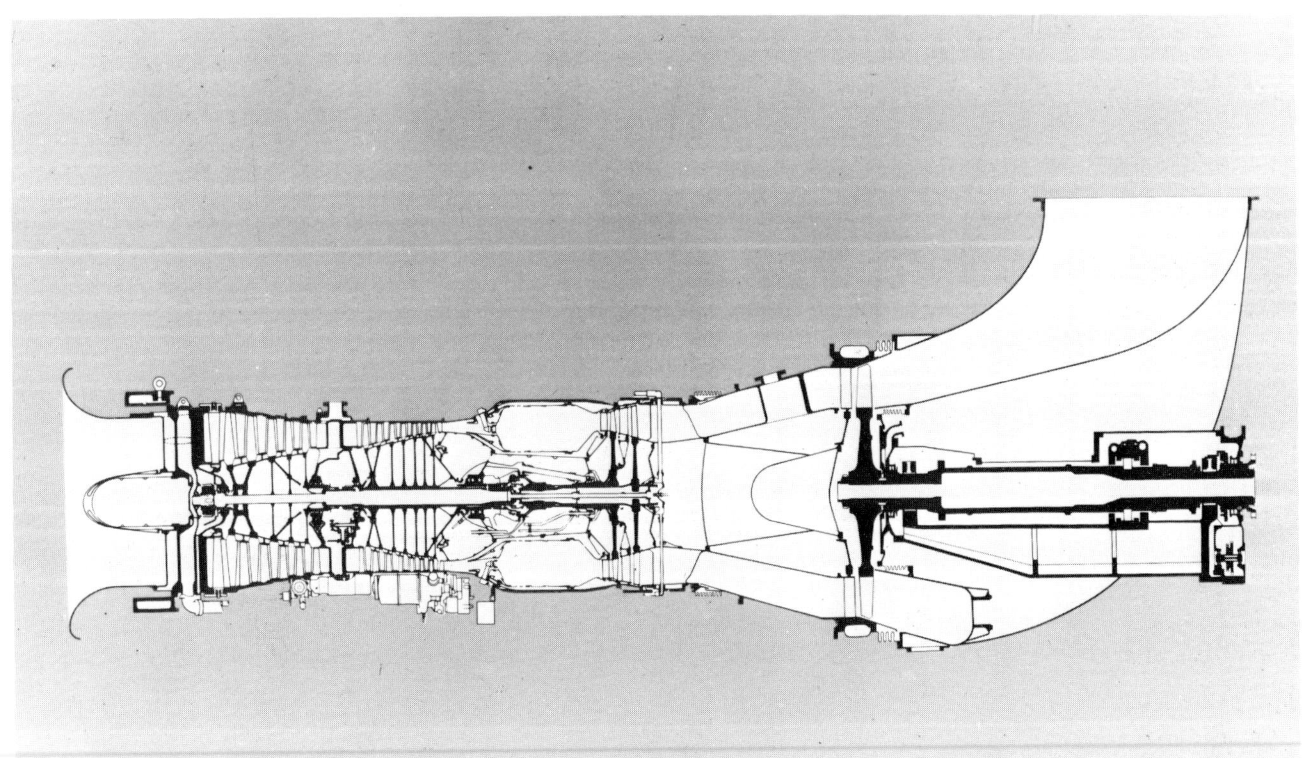

The Marine Olympus. The twin-spool arrangement provides good flexibility over the whole power range, without the need to resort to variable-incidence compressor blades. The single-stage power turbine is heavy by aero standards, but will put up with a good deal of rough treatment.

THE OLYMPUS ARRIVES

The Olympus had been in RAF service for eight years when it was first ordered by the Royal Navy. Looking back from today, it might seem that this engine, now with 25 years of service behind it, is a very ancient piece of equipment for the RN and other navies still to be installing in their newest major warships.

Twenty-five years is a long time in aircraft terms, though not necessarily so by the more conservative standards of the marine world; but it is misleading to use the name 'Olympus' to designate an engine, for it is really a family name. Gas turbine families, like human families, vary widely, from the barren spinster or the still-born baby — such as the Bristol-Siddeley Centaur which we referred to in an earlier chapter — to prolific clans which go on from generation to generation, and scatter their offspring into all walks of life and all parts of the world. Such, for example, are the Rolls-Royce Avon and Spey and, as we noted in the chapter on Hovercraft, the Armstrong-Siddeley Viper — and such is the Olympus.

The first of this clan ran on the test-bed on 16th May, 1950, producing a thrust of 9140lbs, and three years later, fitted in a 'Canberra' bomber, it created a world altitude record of 63,668ft. Two years later the same aircraft, fitted with an uprated version of the Olympus (12,000 lbs thrust), created a new altitude record. In 1956, the Olympus went into service in the Vulcan bomber, where improved turbine blade material enabled the engine to be uprated again, to 13,500lbs thrust. By 1960, a new version of the engine had been developed, the Olympus 201, which powered the Vulcan Mark 2 (17,000lbs thrust).

It was this standard of engine, giving an equivalent shaft power of some 22,000hp, which was to be the starting point for all the subsequent development of the marine and industrial versions of the Olympus over the next 20 years; although power, in the case of the industrial unit, would start at 20,000hp and eventually rise, through successive stages of development, to 37,500hp.

Meanwhile, the aero-engine was continuing to follow its own line of development, and by adding a further stage to the low-pressure compressor, in the Olympus 301, power was again increased in 1963 to 20,000lbs thrust. The following year saw the first flight of the TSR 2 aircraft, which was to be cancelled in a spate of Government cuts six months later, in April 1965. This aircraft was powered by a further development of the Olympus, the 22R — one of the still-born members of the family, but not quite the end of the line, for out of it was developed the power plant for Concorde, the Olympus 593, of 38,0000lbs thrust.

It has to be admitted that the only things which the 593 has in common with its first ancestor are the name and — an important point — the two-spool layout, which was pioneered in the Olympus and has since been used in a number of other engines, including the Rolls-Royce Spey and RB 211 (which has, in fact, three spools in its aero-form). In this arrangement, the compressor is divided into two mechanically independent sections, each driven by its own turbine, with the shaft for the low-pressure section running inside the high-pressure shaft. This system provides a simple and effective solution to the problem, much more acute in high-compression ratio engines, of matching the airflow at the front and rear of the compressor over a wide speed-range, and avoiding the compressor-stall from which, as we have seen, some of the earlier marine gas turbines had suffered. In the 'Olympus' two-spool arrangement, each section of the compressor runs at its optimum speed, as the power output is varied, and there is thus no need for blow-off valves, or the complex variable compressor blading which is the alternative solution to this problem used in some engines.

The Olympus first went into service at sea level as an industrial unit, when a single prototype engine was commissioned by the Central Electricity Generating board in late 1962, at their Hams Hall power station near Birmingham. Proteus-powered generating sets had been in operation since 1959, but their use was confined to the more remote end-of-line parts of the system, and to overseas emergency plant. The CEGB's first essay with the Olympus was a tentative effort, made rather because the gas turbine looked attractive, economically, as a possible peak-load machine, than in response to any real need. In fact, a survey by a market research organisation, carried out in 1961, demonstrated pretty conclusively that there was no market at all for machines of this kind in the power generation field. Fortunately, before this depressing prediction had had any practical effect, both Bristol-Siddeley and the main electrical companies, English Electric and AEI, were deluged with orders from the CEGB, to the tune of 1464 MW in 1962/3 alone, and the power generation market, since that time, has accounted for a total of 10,000 MW from British manufacturers, including over 260 Olympus engines. To be fair to the market researchers, few people had foreseen either the extent to which gas turbine plant would be used to fill gaps at short notice in an expanding electrical system, or the stimulus they would receive from the major power failure in Britain in 1964, and the more serious East Coast blackout in the USA the following year.

Whatever the reasons, the launch of the Olympus as an industrial engine helped to provide a base for sea-level operation, and the development of the industrial version has kept ahead of the marine engine over the years, testing each new step in the growth of power, and providing a wide range of plant for providing modifications or improvements. This is not to say that the marine and industrial engines or applications are the same. The two fields share some common problems, such as operation at sea-level in a nasty environment — coastal stations can pick up as much air-borne salt as a ship in mid-Atlantic — but even in these difficult times, power stations are not normally designed to withstand explosion damage. Nevertheless, there is a good deal of common ground in the design of the gas generators, and both fields profit by an increased scale of production and greater build-up of operating experience.

The first marine Olympus was run in Germany. In 1959, the German navy already had four of their *Köln* class frigates building in Hamburg, at the Stülckenwerft yard (later to become part of Blöhm and Voss) and were committed to using gas turbines; but for later ships they were looking for a more powerful engine than the 13,000hp Brown-Boveri machine in the *Köln* class. Bristol had made a preliminary study of a marine Olympus, with a conservative initial rating of 22,000hp, but in 1959 there was still some reluctance to commit design-effort to a power turbine for such a new-fangled toy. The only immediate solution was to divide the design — not an ideal approach to the problem, but in this case a successful one from a technical point of view. Brown-Boveri were brought in as partners, and after two years of discussion and preliminary design, in May 1962, the German Ministry of Defence awarded a contract to Bristol-Siddeley for the 'marinised' gas generator and the engine control system, and to Brown-Boveri for a new design of two-stage long-life marine power turbine, complete with engine mounting arrangements.

In 1965, after passing its acceptance tests at Mannheim, the engine was delivered to the Navy's technical establishment at Kiel, where a new test bed had been built for extended shore trials. Meanwhile, work was proceeding on the preliminary design of a new class of gas-turbine-powered frigate. The following year, however, the Ministry decided to postpone the building of the new ships, and ordered, instead, three American-built *Charles F. Adams* class destroyers — steam powered. It is ironic that, after this early entry into the marine gas turbine world, the German Navy was not to order any new gas turbine-powered vessels — or, indeed, any new major warships — for almost 20 years,

*The Finnish Navy's corvette **Turunmaa**. Powered by a single centre-line Olympus, with Mercedes cruising diesels on the wing shafts, these ships packed a good weapon load into a small hull, restricted in displacement by Treaty limitation.*

and that the next gas turbine frigates from Blöhm and Voss, after the *Köln* class of 1961, were to be two classes for Nigeria and Argentina, ordered in 1978 and 1979, and powered, at long last, by the Olympus.

The Bristol-Siddeley / Brown-Boveri Olympus did go to sea, but in a slightly different version, in the two 700-ton Finnish corvettes, *Turunmaa* and *Karjalla,* commissioned in 1968. The *Turunmaa* was, in fact, the first ship to operate an Olympus engine at sea, as she started trials at the beginning of 1968, six months before HMS *Exmouth,* the first British ship. These small craft, built by Wärtsila to meet difficult treaty limitations imposed by the Russians, have not attracted as much attention outside Finland as they deserved. Even though the choice of weapon systems was limited to some extent by political considerations, they manage to pack a considerable armament into a small hull; they are well-equipped and well laid out, and they have a genuine maximum speed of over 35 knots, combined with good cruising range. The machinery arrangement is, strictly speaking,

CODAG, with a single Olympus of 22,000hp on the centre-line shaft, and 1200hp Mercedes-Benz diesels on each wing shaft, which can operate independently, to give a cruising speed of 17 knots, or in conjunction with the gas turbine. The gas turbine propeller is of unique design — a shrouded propeller, running in a half-tunnel, which is claimed to give very good efficiency and a low underwater noise-level. One result of the shrouding was that the propeller had to be of fixed pitch, and imposed considerable drag when the turbine was not running, so a third diesel is used on the centre shaft, to drive the big propeller in the cruising condition. Manoeuvring, of course, is carried out on the wing diesels which, in a CODAG arrangement are always running and available to supply astern power in emergency.

The Bristol-Siddeley (or, as we may now say, Rolls-Royce) Marine Olympus started life with the experience already gained in the running of the gas generator at Hams Hall and in a number of subsequent industrial sets, as well as in the

Germany Navy's unit at Kiel, and a number of modifications such as changes in bearings had already been introduced. The work of planning these could be approached with a fair degree of confidence, with the experience already gained on the Marine Proteus. Material changes were comparatively straightforward in the cold area of the engine, where magnesium casings were changed to aluminium (because magnesium reacts violently with salt water) and aluminium compressor-blading was changed, first to stainless steel, and subsequently to titanium. Changes in turbine blade material introduce us to two continuing searches in the aero-engine world: the metallurgist's constant search for better high-temperature alloys and protective coatings, and the development engineer's insatiable thirst for modifications.

Under pressure from the aircraft industry, metallurgy has made tremendous advances over the years, but the introduction of new materials from the laboratory into an engine at sea is necessarily a slow process, since each change can bring not only advances but fresh hazards. Fortunately, much work on one engine can be read across to others, so that, for instance, the build-up of running hours with new turbine nozzles in hovercraft will remove much of the uncertainty when the same material is introduced into an Olympus.

'Modification' is — or ought to be — a synonym for 'improvement', but to say that you are improving something implies that there was something wrong with it in the first place, which most people — and not only engineers — are loath to admit. (In some countries, even the word 'development' is unpopular as it is taken to imply that the device in question is still in the larval, or caterpillar stage, and has not yet reached its destined perfection). Nevertheless, improvements can obviously continue to be made in almost any sophisticated piece of machinery, and in a gas turbine the process is one which only ceases with the final demise of the engine. The lightweight engine or gas-generator lends itself to this process in a way that steam turbines, for instance, do not; for units are removed periodically for inspection and overhaul, and modifications which have already been proven elsewhere can then be introduced without difficulty. To give just one small illustration from another field of how this works

and what it can achieve, the Rolls-Royce Avon engine was first used for pumping natural gas on the Trans-Canada Pipeline in 1964, and at that time it was found necessary to remove engines for repair after some 1500 hours' running, or little more than two months of continuous duty. Today, after successive modification, these same engines are able to run for over 40,000 hours, or five years, before they are removed for overhaul. The rate of improvement is necessarily less dramatic at sea, for an Olympus boost-engine may only accumulate some 500 hours running a year, so if one were able to continue in service for 40,000 hours, we should have to wait until 2060 to enjoy the achievement.

Perhaps the most difficult area in the gas turbine, not in the selection of materials but in design, is the combustion system. This is not surprising when it is appreciated that each Olympus combustion-chamber is burning fuel with energy equivalent to over 10,000 horsepower, with a gas-flow of 30 pounds a second at a temperature of 200°C—all this in a metal cylinder about the size of a vacuum cleaner. This has to be reliable and long-lasting, to cope with rapid variations in power, to stomach at least a proportion of undesirable elements such as sodium and sulphur, and to produce completely smoke-free combustion throughout its power-range. The first Olympus engines were by no means perfect as regards smoke, and it was only after some years of painstaking improvement that the Royal Navy could claim a completely clear funnel at all speeds.

The power turbine, operating at more modest temperatures and pressures, presented fewer development problems. Whereas the industrial power turbine, built to drive a 50- or 60-cycle electric generator, had to run at 3000/3600 rpm, the marine version, which would in any case drive through conventional reduction gearing, could be designed to run at higher speed, which meant that good efficiency could be obtained from a single-stage machine, with a consequent saving in weight, space and cost. The design chosen for the Royal Navy's Marine Olympus used a single stage running at 5660 rpm, with stout, wide-chord blades, and overhung from a rear-facing shaft, so that the plain bearings and thrust collar were clear of the hot gas-stream and accessible for inspection. This arrange-

*H.M.S. **Exmouth**, the world's first major warship to rely entirely on gas turbine propulsion, with one Olympus and two Proteus cruising engines. One of the 'Blackwood' class steam frigates, she was converted to gas turbine propulsion to provide a floating "test-bed" for the Olympus, but she carried on with her normal Fishery Protection duties after trials were completed.*

*The Vosper Mark V **Saam**, first of a class of four Iranian destroyers. Their twin-Olympus CODOG machinery gave them a dazzling top speed of 39 knots.*

ment has become standard in virtually all the large British gas turbines, and as it has given no trouble, it will have little place in our history.

There remained the question of shock-resistance, which we have already touched on in an earlier chapter. As far as the power-turbine was concerned, this presented no problem. The gas-generator, as built to aero-standards, was less robust. However, most of an aero-engine's components — the discs and blades and shafting — will stand up to very high shock-loads, and it had already been calculated that increased bearing sizes and strengthened casings would go some way to meeting naval requirements in the more vulnerable areas. When the modified gas generator was mounted with the comparatively heavy power turbine on a deep base plate, the complete unit was calculated to withstand accelerations up to 30G, a figure later confirmed by shock-barge trials carried out by the Naval Construction Research Establishment at Rosyth.

Shock-barge trials are very spectacular: whether they actually prove anything is a debatable point, for the whole question of damage from underwater shock is a difficult one. To quote such figures as 30G can be misleading, as the stresses set up depend on the duration, as well as the intensity, of the high G-load. Obviously no marine engine is going to be subjected to sustained accelerations, as is the case in aircraft when manoeuvring — 30G sustained for even half a second would leave an engine 120 feet up and travelling at 300 miles an hour. It is equally clear that, if the period of shock is sufficiently short, enormously high accelerations can be tolerated, and in fact shock barge trials have recorded transient readings as high as 300G without damaging an engine. In practice, the Navy's requirements call for engines to withstand considerably higher shock-loads than 30G for short periods; but with the standard which had been achieved in the Marine Olympus, the balance could be made up without difficulty by fitting the complete unit with shock-attenuating mountings.

The first Rolls-Royce Marine Olympus started running on the test-bed in August 1966 and continued, with intervals for modifications, until mid-1968, when the first seagoing unit went into service. When the decision to develop the Olympus had been taken by the Ministry of Defence in 1964, the intention was to use it, initially, as a boost engine in a COSAG arrangement for the new Type 82 Guided Missile Destroyers, the first of which, HMS *Bristol,* was ordered in 1966, soon after the start of the Olympus shore-trials. In order to get early sea-experience with the new engine, ahead of its commissioning in HMS *Bristol,* it was decided to convert an existing steam-powered *Blackwood* class frigate, HMS *Exmouth,* and it was in this ship that Olympus sea trials started in 1968.

By this time, however, a number of fresh decisions had been made. In 1967 the Ministry of Defence finally decided to use gas turbine propulsion for a range of new frigates and destroyers. Initial plans were for a class of 3500-ton destroyer and a smaller design of frigate. In 1968, as we shall see later, the Ministry took the unusual step of awarding a design contract for the frigate to Vosper-Thornycroft and Yarrow. Both firms could claim full competence for the task, for, apart from a long tradition of small warship building, both had recently secured overseas orders for modern gas turbine-powered warships. Vosper had received a contract in 1966 for a class of four small destroyers for the Iranian Navy, which were to be equipped with Olympus CODOG machinery, while Yarrow had a contract six months earlier from the Malaysian Navy for a frigate, also with CODOG machinery.

The Yarrow design was for a ship of 1600 tons with a conventional general purpose armament, including a 4.5 inch gun, Seacat SAM, an A/S mortar, and the ability to land a helicopter. The machinery consisted of a single Olympus, giving a speed of 27knots, and a Crossley-Pielstick diesel which gave a cruising speed of 16 knots. The conventional transmission for such an arrangement, with the single boost and cruise engines mounted on the centreline, would have been with a single shaft. This was the propulsion system used in HMS *Exmouth* and other *Blackwood* class frigates. It is certainly the cheapest arrangement, and a single shaft with two large gas turbines was chosen ten years later in the American *Oliver Hazard Perry* class frigates, largely for economy reasons. Yarrow, however, adopted a twin-shaft arrangement, with a gas turbine and diesel both feeding through clutches into a

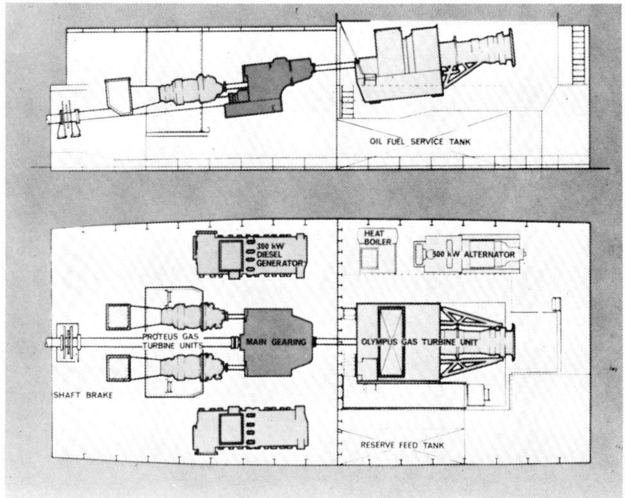

ABOVE: *The Malaysian Navy's Rahmat, built by Yarrow in 1971, with single Olympus CODOG machinery. With her sister ship, the Thai **Makut Rajakumarn**, she was the first of a long line of small frigates or corvettes using a single Olympus with various combinations of diesel cruising engines.*

BELOW: *Exmouth machinery layout. The Royal Navy's standard cruising engine, the Tyne, was not yet available when Exmouth was converted, so the well-tried Proteus was pressed into service.*

common gearbox, which provided a split output. This avoided the vulnerability of the single-screw ship to propeller damage, and still gave good propulsive efficiency, with reduced draught and improved manoeuvrability. Single-shaft ships have always been unpopular with their drivers because of their lack of manoeuvrability, and although the controllable-pitch (CP) propellers used in the Yarrow frigate (and in nearly all gas turbine-powered ships since then) may not be so effective for harbour manoeuvring as reversible propellers, they are a great improvement on the single screw.

The Yarrow type of machinery arrangement was to be used in a long succession of small frigates and corvettes — in the two frigates built by Yarrow themselves for the Malaysian and Thai Navies, in the Belgian Navy's *Wielingen* class, in the *Fatahillah* class corvettes built in Holland for the Indonesian Navy, and in the two

Fatahillah, first of a class of corvette built by Wilton-Fijenoord for the Indonesian Navy. She uses a machinery arrangement similar to that in the earlier Yarrow frigates, with a single Olympus driving twin propellor shafts, and with two MTU diesels providing cruise propulsion.

Uniao, one of the 3500 ton Brazilian 'Niteroi' class destroyers. Of Vosper design, the last two of the class, **Uniao** and **Independencia,** were built in the Brazilian Navy's yard in Rio de Janeiro. Powered by two Olympus and Brazilian-built MTU diesels.

training ships built by the Yugoslavs for Indonesia and Iraq. In each of these classes, apart from the Yarrow boats, the Olympus is used with two cruising diesels, so that, although a single main reduction gearbox is employed, the diesels can be isolated, each driving its own propeller shaft. In the most recent vessels of this type, the Japanese *Ishikari* class frigates, a single diesel and Olympus are used, as in the original Yarrow design.

The Vosper *Saam* class destroyers built for the Iranian Navy were of about the same displacement and dimensions as the Yarrow frigates, but they were equipped with two Olympus engines, which gave them a sparkling high-speed performance, taking full advantage of the saving, both in weight and space, and in engine-room complement, which was now being made possible by the gas turbine. These ships, using two independent shaft lines of machinery with Paxman diesels, cruised at 18 knots and had a maximum speed of no less than 39 knots in the half-fuel condition. In addition to a 4.5inch gun, Seacat SAM, and A/S mortar, they were equipped with Contraves Sea Killer missiles, and a slightly larger version, the Mark 7 frigate *Dat Assawari,* which was ordered by the Libyan Navy in 1968, now carries Otomat missiles and two triple A/S torpedo tubes.

Vospers were later to apply the same design philosophy, with the same undoubted sex-appeal, to a much larger class of destroyer for the Brazilian Navy. These were the six Mark 10 class, of 3500 tons, carrying a Lynx helicopter, and a considerable armament including a 4.5inch gun, Seacat, Exocet and Ikara missiles, as well as ASW rocket-launchers and torpedo-tubes. To provide higher cruising speed the machinery included four Brazilian-manufactured MAN diesels, totalling 15,760hp, which gave a speed of some 22 knots, while the Olympus gas turbines, by then uprated to 28,000hp, gave a maximum speed of 30 knots. These ships properly belong to a later part of our story, as the first of class, *Niteroi,* was not commissioned until 1976.

To return to 1968: while the design of the new RN ships proceeded, HMS *Exmouth* was being converted for Olympus sea trials. She was not an ideal trials vehicle as her existing steam machinery developed only 15,000hp. As the same shaft line was maintained as well as some elements in the existing gearbox, the Olympus was also limited to 15,000hp, although the mark of engine installed, the TM1A, had a maximum power of 24,000hp. She did, however, provide a quick means of getting the engine to sea and evaluating gas turbine propulsion in an operational ship, so that control and operation, manoeuvring, maintenance, noise, and seagoing ability could be assessed, and any necessary alterations made in installation design or operating and maintenance procedures, before the new engines were introduced generally in the Fleet.

As is always the case with the Navy, money

was tight, and the *Exmouth* conversion could be carried out for a relatively low cost, during a period already earmarked for refit. After trials, the ship could return to her normal operational role and continue with fishery protection duties, so that she would only have to be spared from active service for a relatively short period.

The decision taken by the Ministry of Defence in 1967 involved the use of gas turbines for cruise propulsion as well as high speed, but the engine chosen for the cruise role, the Rolls-Royce Tyne, was not yet available for sea trials. It was nevertheless necessary, in purely practical terms as well as to assist sea evaluation, to use some sort of cruising gas turbines in *Exmouth,* and the Proteus was once more pressed into service, two units being mounted on the after side of the main gear box. Proteus and Olympus drove through SSS clutches in a COGOG arrangement, and a controllable-pitch propeller was used for manoeuvring, so that, in terms of general layout and control, the machinery would follow the pattern planned for the new frigates and destroyers.

On 5th June, 1968, *Exmouth* started sea trials — the world's first major warship to have all-gas turbine propulsion. Just nine days later, after 64 hours' running, the Olympus suffered a major failure. The entire first row of LP compressor blades was lost, and trials were abruptly halted.

This was a major setback to a programme on which so much depended, the more so since it seemed quite inexplicable. There was no question of the compressor blades not being good for far longer running under normal circumstances. By this time there was already a considerable number of industrial Olympus engines in service, and these used the same stainless steel compressor blades. Over 27,000 hours running had been logged, and many individual engines had totalled over 1000 hours in electrical peaking duty — much more arduous than marine operation — without any sign of distress. It had been shown, by detailed testing, that the blades were not exposed to any abnormal stress in other installations, and there was no significant corrosion in any of the broken blades. By routine detective work, the metallurgists were able to establish that the damage had not been caused by any foreign object entering the intake, but that the blades had failed from a quite unprecedented level of vibration.

There was no other source of vibration in the ship which was capable of doing the damage — this could be proved by running at sea, under the same conditions as before, on the Proteus engines. The only possible conclusion seemed to be that the air flow in the intake was being grossly disturbed by some feature in *Exmouth's* downtake trunking.

Conversions of existing ships are always liable to pose problems, since the installation of new equipment frequently conflicts with the layout of existing plant and one or the other — or both — have then to be modified. To continue the detective work on the *Exmouth* failure, it was necessary to have an exact plan of the downtake, as actually built into the ship. Once this had been secured, by on-the-spot measurement, models were made and tested, including a fifth-scale model for quick measurement of air-flow patterns and pressures, a water-analogy model, in which, with water taking the place of air, the flow pattern could actually be seen, and, finally, an exact full-scale replica for installation on the engine test-bed.

The result was dramatic. The asymmetrical design of the downtake, necessitated by having to insinuate it into an existing hull, had caused a breakaway in the air flow on one side, at the bottom of the downtake trunk, producing an intense vortex centred on one side of the engine intake, which meant, in effect, that each compressor blade, as it reached the 'four-o'clock' position in its rotation, was passing through a point where the air flow had virtually stopped.

This is a situation which might occur in any ship installation. Engineers legitimately have the major say in the layout of machinery spaces, but they cannot always control in detail what is done by the ship constructors in more remote areas, and engine downtakes have to run through the depth of the ship. Even if the constructors can be controlled, action damage may upset the flow of air in the downtake. To insure against this, it was concluded that the air-flow, whatever vicissitudes it might have suffered on the way down, must be smoothed out within the engine mounting, so that the gas turbine would be protected alike from the worst efforts of both the ship designer and the enemy.

Within just four months from the failure, the engine had been removed and repaired, the models made, the tests done, the calculations completed, the solution found and the installation altered, so that trials were able to start again on 15th October.

This time there was no failure, and the cascaded duct in front of the engine air intake, which effectively cured the *Exmouth* trouble, has been a standard feature in all later engines.

It may seem that a disproportionate amount of space has been devoted to this one failure; but it serves to illustrate several lessons. In the first place, it does demonstrate the value of sea-experience, for the type of trouble met with here, though it might occur in any ship installation, would not have shown up in shore-testing, unless a complete replica of the ship installation has been built on the test bed — a practice now largely followed, in order to bridge the gap, as far as possible, between initial engine testing and sea trials. In the second place, it points a moral: that an engine — or any other piece of equipment — needs, as far as possible, to be insulated against any sort of hostile external influences, even those which we cannot entirely calculate or foresee. Third, it is worthy of remark that a major investigation of this kind could be planned, carried through, and acted upon in such a short time; but the speed with which development engineers went to work, and the resources which could be brought to bear on the problem, would not themselves have got the ship back to sea again so quickly, had the engine not been designed for the rapid repair or replacement of parts.

Perhaps the final reason for dwelling on this incident is that it was, in fact, the only serious failure which occurred in the *Exmouth* engines. For the most part the trials and the subsequent running went extremely smoothly and without incident. This is not at all to say that gas turbines, even the best, never go wrong. Any type of machine can, and does, go wrong. Gas turbines are no exception, and the very business of striving for improvement brings one into an area of uncertainty; but experience helps to pin down the uncertainties, and with a solid background on which to build, the marine gas turbine has, on the whole, proved remarkably trouble-free and easy to maintain. The only complaint heard from the Engine-room Artificers in *Exmouth* was that they hadn't enough work to do.

When she had finished her period of gas turbine trials, *Exmouth* returned to the fleet, and continued in service until 1979 when, by then 22 years old, she was finally paid off. But long before then, she and her successors had demonstrated beyond doubt the advantages of gas turbine propulsion, and the day of the steamer was over.

COGOG

n the early 1960s, when gas turbines were first going to sea in major warships, there was a growing move to get rid of the complication of steam machinery and use diesel engines for cruise propulsion. Coupled with gas turbine boost engines, the diesel provided a very economical solution to the perennial warship-propulsion problem of combining economic and reliable cruise power, and a far higher maximum power which was hardly ever used, in a hull which was always limited for weight and space.

The Italians and the Germans both introduced CODAG frigates; Vosper used the Paxman diesel, Yarrows the Pielstick, in CODOG machinery, and the British Ministry of Defence, though not yet committed to abandoning steam, had its own ASR-1 diesel in service and was interested in the development of the new Ruston diesel, up to 8000hp. The combination of gas turbine and diesel was, and still is, a very good arrangement in many ways.

The idea of using gas turbines for cruise propulsion was not new. The English Electric EL 60A had been designed as a cruise engine; the Rolls-Royce RM 60 had been envisaged in that role, and the all-gas turbine ship had long been seen by many people as the ultimate goal. Thinking in the earlier days, however, symbolized by the EL 60A and RM 60, had tended to assume that any cruise gas turbine would have to use a complicated regenerative cycle to give sufficiently good fuel-economy at low power. This assumption did not help the case for the gas turbine, since the main advantage of gas turbines over diesels lay in the major reduction of onboard maintenance, and the consequent reduced manning and increased availability of the ship. If the gas turbines were going to incorporate heat-exchangers, not only would they be heavy and bulky, but they would also require increased maintenance, and it would scarcely be possible to remove them from the ship for servicing.

It could be claimed for the gas turbine that it was lighter than the diesel — but this did not apply to regenerative turbines. It might be claimed that the gas turbine saved space — but this, again, did not apply to large regenerative machines, and in any case it was largely off-set by the diesel's smaller intake and exhaust. The diesel's main advantage was in its low fuel-consumption, and the best regenerative gas turbine could not approach it in this respect, while earlier simple-cycle gas turbines were hopelessly uncompetitive, with a specific fuel consumption around 0.8lbs/hpr-hr against the diesel's 0.4. Quite apart from the expense — and the Navy has always zealously economised on fuel, even in the halcyon pre-1974 days, when it only represented about 4 per cent of a ship's total running cost — the extra fuel-stowage required for gas turbine cruising would more than offset the saving in machinery weight which was still one of the simple-cycle gas turbine's advantages.

The gas turbine, however, was making great strides. By 1966, when the Canadian Navy were designing their *Iroquois* class destroyers, they were prepared to accept a choice between the Proteus, with a specific fuel-consumption of 0.58lbs/hp-hr, and the Pratt and Whitney FT12, with a similar efficiency, and to opt finally for all-gas turbine propulsion; but this was still very much a marginal decision. By this time, however, another potential marine gas turbine was appearing on the scene, in the form of the 4000hp Rolls-Royce Tyne.

The Tyne, in fact, first went to sea in 1966, though the version of the engine which was used at that time was virtually the same as the turbo-prop aero-engine, which had started passenger service in the Vanguard in 1960.

This was developed by Rolls-Royce as a 'second generation' turbo-prop engine, but by the time it went into service, the propeller-driven passenger aircraft was already being ousted by the jet — prematurely, as many people would maintain today. Nevertheless, 44 Vanguard aircraft were built, and the Tyne, outstandingly economical at that time — and still, today, the most efficient engine in its power-range — was also used in a number of other aircraft, including the Canadair CL 44, the Breguet Atlantic, and the Transall C-160. As a result of this, when it came to be considered for marine use in 1966 it had the advantage of having already built up over 3.8 million hours in aero applications. This would certainly not guarantee success in the very different world of ship propulsion, but it did give good assurance that at least the aerodynamic and mechanical bugs should have been largely eliminated. Thermal and metallurgical bugs were another matter, and would have to be submitted to the full marine treatment, though this was now becoming better understood, and its results more predictable.

In one respect, the Tyne required more extensive modification than its predecessor, the Proteus. Like the Olympus, it was a two-spool engine, with separate high-pressure and low-pressure compressors running on concentric shafts and each driven by its own turbine; but as the Tyne was a turbo-prop engine rather than a jet, the low-pressure compressor shaft was also the output shaft, and the low-pressure turbine provided the power both for its own compressor and for the aircraft propeller. This meant that, for marine propulsion, the engine had not the same operational flexibility as the Proteus, as the output shaft speed was tied to that of the low-pressure compressor.

In the first marine Tyne, this could be tolerated, as it was applied initially in the Grumman hydrofoil, USS *Flagstaff* — rival of the Proteus-powered USS *Tucumcari* which we discussed in an earlier chapter. Two versions of this craft were produced; the military one for the US Navy, and a civilian passenger ferry, the *Dolphin,* which was built by Blöhm and Voss, and operated a service in the Canary Islands. Unlike the water-jet propelled *Tucumcari, Flagstaff* and *Dolphin* used conventional propellers, driven by a vertical shaft through bevel gears, and the Tyne's lack of flexibility in matching speed and power requirements could be circumvented without difficulty by using a controllable-pitch propeller, which enabled thrust to be

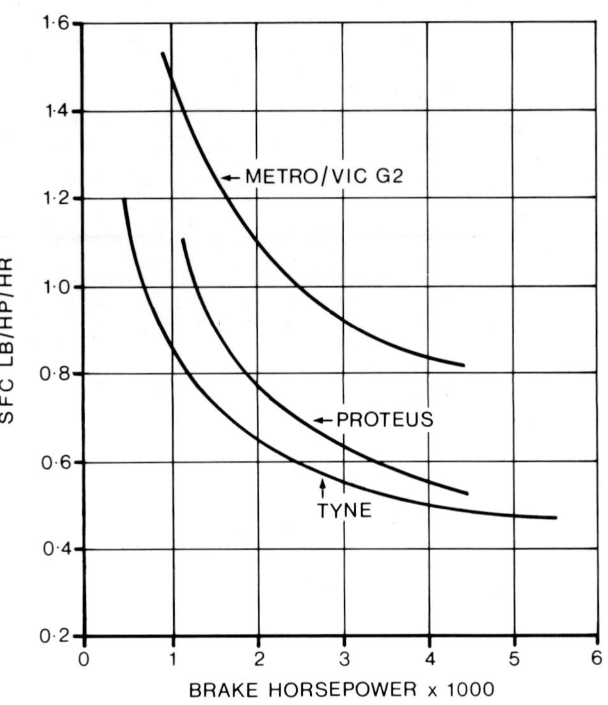

Progress in engine efficiency. From the G.2, in 1953, to the latest version of the Tyne, in 1968, the specific fuel consumption of the small gas turbine has come down from 0.82 lbs/hp-hr to 0.49 — an increase in efficiency of nearly 70%.

varied as required for the very different conditions of take-off and high-speed foil-borne operation.

For operation as a cruise engine in a big ship, however, it was essential that the engine should have the flexibility of the free-power turbine, and radical alterations were needed. As it happened, the low-pressure turbine of the aero-Tyne had three stages, and it was found that, virtually without any aerodynamic re-design, this turbine could be chopped into two separate sections, with the first stage driving the low-pressure compressor and the two remaining stages driving independently on a new shaft towards the rear of the engine, to form the power turbine. The more complex two-spool arrangement of the Tyne, compared with an engine such as the Proteus, reflected the advance made during the years which had elapsed between the initial design of the two engines. The earliest marine gas turbines such as the G.2 had had a compression-ratio of 4:1; in the Proteus this was increased to 7:1, but in the Tyne it was 12:1. Operating temperatures had also increased: in the G.2, the gas temperature at entry

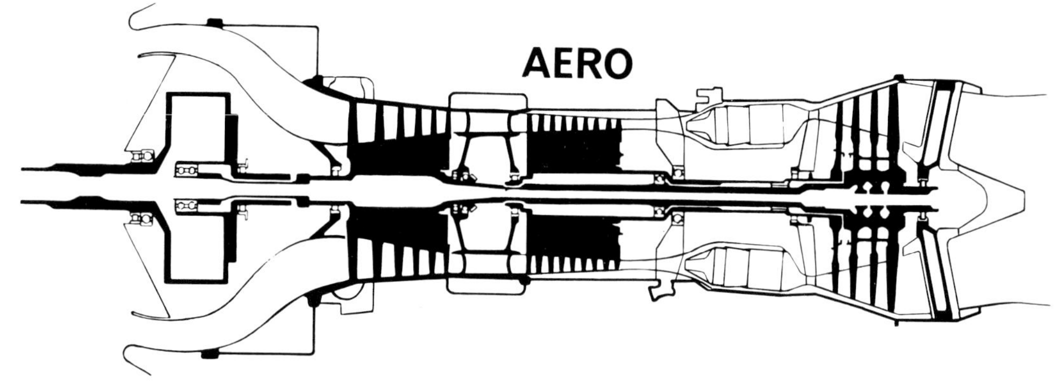

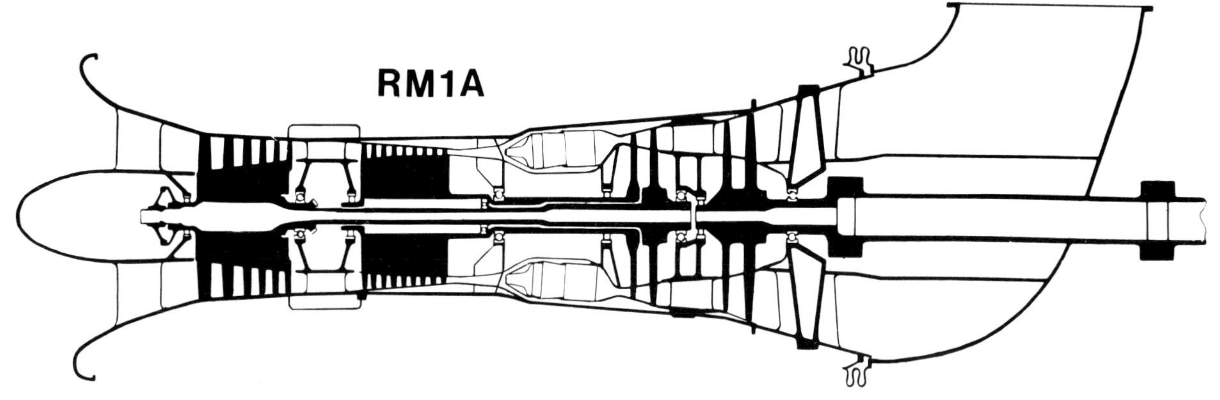

The standard cruising gas turbine in the Royal Navy, and in the Netherlands, Japanese, and Argentine Navies, the marine Tyne still today has an unrivalled efficiency for an engine in this power range.

to the turbine was 1070°K; in the Proteus, 1185°K. This, however, was for the full power of the engine when it was being used for high speed operation in its fast patrol boat application. For regular passenger service in the hovercraft, a lower power-limit was imposed, and the same policy was applied to the Tyne when it first went into marine service as a big-ship cruise engine in 1974. The gas temperature at entry to the turbine was limited to 1155°K, which was the same figure as that for the original 24,000hp Marine Olympus. It represented what was then regarded as a safe ceiling for turbine entry-temperature in a marine atmosphere, always accepting that, while the demand for power would inevitably increase, the first essentials were reliability and long life. Later versions of both the Olympus and the Tyne would run at higher temperatures, through the introduction or extension of turbine blade cooling, and the use of improved materials, but this stage had not yet been reached when the first COGOG ships were designed.

The increases in temperature and compression ratio brought a corresponding improvement in efficiency. While the G.2 had a specific fuel consumption of 0.82 lbs/hp-hr, and the Proteus 0.58, the figure for the Tyne was 0.49 — and here we are approaching the 0.4lbs/hp/hr which the rival diesel can offer, and the small gas turbine's one outstanding drawback as a cruise engine is being reduced to manageable proportions — to a level at which the simple-cycle gas turbine's advantages of low maintenance and high availability can tip the balance in its favour.

Admittedly, the good efficiency of a simple-cycle engine such as the Tyne falls off considerably as power is reduced. Thus at half-power, just over 2000hp, the fuel consumption of the Tyne had increased from 0.49 to over 0.60 lbs/hp-hr — 50 per cent greater than an equivalent diesel engine. This seems to be a considerable drawback to using gas turbine cruising engines. Nevertheless, the reduced efficiency of a Tyne at half power is not quite so damaging as it may sound. In a typical modern destroyer or frigate, in which 50,000hp would produce a speed of some 30 knots, two Tynes at half power give a speed of about 14 knots. But this is not quite the end of the story. In the first place, there is in fact little difference in the fuel consumption in tons per mile over a range of speeds, and 16 knots is probably almost as economical as 14 knots. This is admittedly partly due to the gas turbine's efficiency falling off at low power, but it is also affected by the fact that the 'hotel load' — the auxiliary power required for electronics, ship's services, and domestic purposes — has increased greatly over the years, and now accounts for a considerable proportion of the total fuel consumption at sea. This auxiliary fuel consumption is, of course, at a fixed rate per hour, and therefore at a rate per mile which increases as speed is reduced. The net result is that there is little benefit in running at very low speeds with gas turbine plant or with diesels. However, if maximum economy with COGOG plant is essential on a long passage, it is perfectly feasible — and, indeed, is today normal practice — to run on only one cruise engine, allowing the other shaft to trail. This means that, for a speed of 14 knots, our single Tyne is producing nearly the maximum power, and running at much high efficiency, in fact with a total ship fuel-consumption between 15 and 20 per cent above on equivalent diesel-powered vessel — and a good deal lower than with contemporary steam plant.

It may be felt that we are dwelling too much on fuel-economy, but this has always been a paramount concern in the Royal Navy, and only very powerful reasons have ever been able to justify using other than the most economical plant, procedure, or speed. Indeed, the rules restricting speed seem to have been more stringent in earlier times, before the Second World War, than they are today. In those days any commanding officer who exceeded 'economical speed' without permission was expected to give very compelling 'reasons in writing', which meant that, for most of the time, the whole Fleet made its stately progress through the oceans of the world at 12 knots.

This is, perhaps, not surprising, for except during a war, naval spending since Pepys's time has nearly always been under political attack — even if not always quite so viciously as at present — and fuel is one of the very few items on which day-to-day savings can be made. It is difficult to switch off a current building programme, even though new ships may be "mothballed" as soon as they are completed; it is impracticable to lay off sailors, and even captains are not nowadays placed on half-pay, as they were wont to be in earlier times; reducing pay had an unfortunate effect when it was last tried at Invergordon. Fuel can be saved much more readily, however, by steaming at economical speed or merely by staying in harbour; but more lastingly, and with less harm to the effectiveness of the Fleet, by using more efficient machinery.

The fact that we have reduced the gap between diesel and gas turbine efficiency does not mean that the gas turbine's disadvantage has disappeared. It entails two penalties, which have to be weighed with the other pros and cons in the diesel-versus-gas turbine debate. We have seen that, in the matter of cost, the gas turbine ship uses 15 — 20 per cent more fuel when cruising than does its diesel counterpart. This is at its best; under some conditions, the disparity will be higher. On the other hand, at high speed — which is admittedly seldom used — both the CODOG and the COGOG ship are running on boost engines, so the choice of the cruising engine makes no difference to fuel consumption. The overall difference, in terms of annual fuel cost, depends on the operating pattern — in other words, the percentage of total steaming time which is spent at different speeds. For a typical frigate operating-pattern, annual fuel-cost difference works out at around 10 per cent, and since fuel-cost represents today some 15 per cent of total ship-operating cost, the use of cruising gas turbines in a COGOG arrangement is likely to increase this total cost by about 1.5 per cent.

The other penalty incurred by higher fuel-consumption is the need to increase fuel-stowage in order to maintain endurance. The requirements for endurance vary, but to illustrate the point, we can take a typical figure, for a frigate or destroyer, of 5000 miles at 18knots. For this speed our average ship will need 10,000hp, and taking 0.4lbs/hp-hr for the diesel and 0.48 for the later (5000hp) Tyne, we can see that, with the gas turbine, fuel-consumption increases from 500 tons to 600 tons, or other things being equal, our deep displacement goes up from, say, 3500 tons to 3600 tons. This is unlikely to alter performance significantly, or have any adverse effect on stability, but space must be found for the additional stowage.

Against these two space and cost penalties we have to set the reduced manning costs and higher availability of the gas turbine ship.

When the diesel cruising engine and the gas turbine were being evaluated in the 1960s it was noted that the gas turbine had one other advantage over the diesel, which was to grow increasingly important as submarine detection equipment became more sensitive and more demanding. This was the very low level of underwater noise which it generated, by comparison with the diesel.

Gas turbines, as we noted earlier, had — and still have in many people's minds — a reputation for being extremely noisy. This is quite unfair. The shattering roar of the jet engine comes largely from its exhaust, which, of course, also provides the engine's power in the form of thrust. In the marine gas turbine, however, the energy in the exhaust from the 'jet' is used to drive the power turbine, and so provide shaft power rather than thrust. After giving up their power the exhaust gases become very much more manageable, and while they still need to be treated with respect in the design of the uptakes and funnels, their noise is reduced to a relatively subdued rumble — so much so that, although silencers were installed in the uptakes of earlier gas turbine frigates and destroyers, these have now been largely dispensed with, as serving no useful purpose, since at sea any noise coming from an unsilenced funnel is swallowed up in the general background sound of wind and water and ship's fans.

The jet engine does also produce a considerable amount of noise at the air intake. This has been greatly reduced in recent years, not only by the introduction of turbo-fan engines, but also by improvements in the design of blading and struts at the front of the engine, to cut out much of the objectionable high-pitched scream which is characteristic of this end of the machine. The aircraft, however, is at a fairly obvious disadvantage compared to a ship in that it can hardly fly around with large silencers in front of its jet engines. In a ship, on the other hand, air has to be ducted from the upper deck through a considerable length of vertical trunking, and it is a matter of no great difficulty to install silencing 'splitters' in the downtakes to achieve any degree of intake-silencing that is required. Some noise is radiated from the engine carcase itself, but this again can be dealt with comparatively easily. Unlike the diesel, the gas turbine radiates noise that is predominantly high-frequency, which is relatively easy to absorb by using conventional sound-insulation material. This is done in the ship's engine room by housing the complete unit in a steel enclosure clad with insulating panels. This system also provides heat-insulation and a protective shield for the engine, as well as affording a means of isolating the ship's machinery spaces from any radio-active contamination in the intake air, which might leak from the engine casings. We are thus equipped with a system of soundproofing which can be varied to suit any requirement.

For a ship installation the only need is to make it practicable for men to work in the engine-room without distress or inconvenience. Where the same technique is used for industrial plant, the degree of silencing required — for instance, on a big gas-pumping plant in a rural area — may be very much greater. There is no technical problem in reducing gas turbine noise down to almost any level. Additional silencing will obviously add something to the cost of the equipment, so it will not be indulged in beyond the extent necessary to ensure good efficiency and a comfortable working environment.

All these forms of silencing, however, are largely a matter of convenience and habitability: of much more vital importance in anti-submarine operations is the elimination of underwater noise, transmitted from machinery through the ship's hull. Great efforts have to be

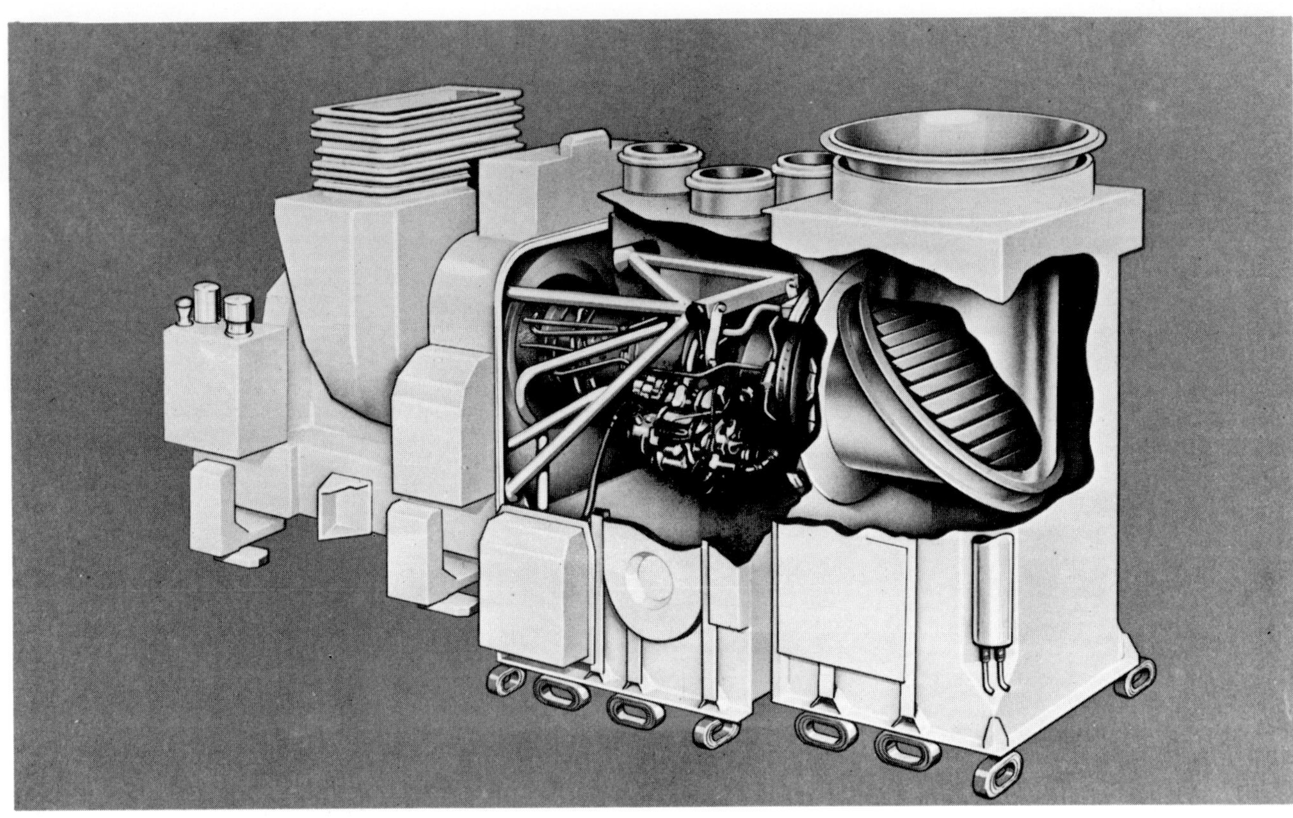

For ship installation, the marine Tyne is mounted on a rigid base-plate with a heat and noise-insulating enclosure and ducted intake, in similar fashion to the Olympus. The Tyne also incorporates its own primary reduction gear, to bring the output shaft speed down to a suitable level for ship's main gearing.

H.M.S. **Newcastle**, a 'Sheffield' class destroyer of 3500 tons. It was for these ships that the standard Olympus-Tyne machinery package was first developed.

made to deal with every possible underwater noise source in a modern ASW frigate or destroyer, and the diesel, with its sharply accented, characteristic low-frequency beat, is particularly difficult to deal with. Modern high-speed diesels, with double noise-insulating mountings, have improved greatly; but the sensitivity of underwater detection equipment has also increased, so that it seems unlikely that diesel propulsion will ever be entirely satisfactory for anti-submarine operations.

By contrast, the gas turbine, with its predominantly high-frequency noise, very low vibration levels, and relatively light weight, can be insulated from the ship's hull without difficulty, and this is one compelling reason why most of the world's major ocean-going navies have adopted some form of all-gas turbine propulsion.

Of course, with combined gas turbine and diesel machinery, it is perfectly possible to run on boost gas turbines while engaged in anti-submarine operations, and in many smaller navies, where ships are used largely for coastal and off-shore 'policing', fishery protection, and so forth, with anti-submarine warfare as a secondary role, this may be quite acceptable. But it is certainly not economical if anti-submarine operations are to last any length of time.

It is impossible in a few pages to argue all the pros and cons of diesels and gas turbines, because the balance of advantage is affected by so many factors, in particular by the ship's intended role and expected operating-pattern. If we say that the gas turbine has the advantages of low maintenance, with reduced manning and higher availability, plus low underwater noise, against the diesel's reduced fuel-consumption, we need to add two other reasons for the selection of diesel cruising engines in some navies. The first is a natural inclination, particularly in smaller navies, to choose 'the devil you know'. This is not altered by the fact that 'the devil you don't' turns out, in the long run, to be easier to maintain and operate. Moreover, it can, of course, be based on sound reasoning if maintenance and training facilities are already available for a suitable diesel. The second reason for choosing a diesel may be that a locally manufactured engine is available, and this can certainly in some cases be a sufficiently strong factor to override technical arguments.

One other gas turbine advantage needs to be stressed, because it was a dominant factor in the final decision of the Royal Navy, though it is seldom given sufficient weight in diesel-v- gas turbine discussions. This is the improvement which the gas turbine effects in the quality of life on board for engine room personnel. Good working conditions are not merely desirable on social grounds; they make an undoubted contribution to the efficiency of a ship, besides helping recruiting, and the improvement of recent years is nowhere better exemplified than in the shift from steam or diesel to gas turbine propulsion. This has made an astonishing difference, not only to the amount of maintenance and watchkeeping required, but to the type of work involved and the conditions — of noise, heat, dirt and space — under which it has to be carried out. It is a fact that few engineers who have served in gas turbine ships have any wish to go back to the old steam and diesel days.

As far as the Royal Navy was concerned, the long debate was finally resolved in 1967; the decision was taken to use all-gas turbine propulsion in all new classes of frigates and destroyers (the choice of gas turbines for the new aircraft carriers was to follow two years later), and in August the Ministry of Defence placed a contract with Rolls-Royce for the full development of the Marine Tyne.

The Royal Netherlands Navy's destroyer **Tromp,** first fruit of a major Dutch new building programme and, at 5400 tons, the largest class of ship to use the standard Olympus-Tyne COGOG machinery.

THE GAS TURBINE FLEET

The transition from steam to gas in the navies of the world covers a period of some 30 years. From the early days, after the close of the war up to 1960, almost all the gas turbine activity was experimental or, later in this period, confined to first essays in a few isolated craft.

During the 1960s the gas turbine was gaining recognition as an established type of prime mover, but although the philosophy was changing, there was still remarkably little ironmongery to demonstrate the faith of the philosophers. By 1970 the RN did have 15 COSAG frigates and destroyers at sea, but only one experimental all-gas ship, the *Exmouth.* The Germans had the *Köln* class, the Italians were experimenting with some conversions, the Finns had built their two corvettes, and the Danes had completed two CODOG frigates. The United States Navy had one gas turbine transport on charter, though the US Coastguard had commissioned a number of *Hamilton* class CODOG cutters, using the Pratt and Whitney FT4.

From 1970 onwards, however, we are no longer talking philosophy. The main decision has been made, and the argument is now about the relative merits of different types of gas turbine, alternative machinery arrangements, and varying ways of meeting changing operational needs.

For the Royal Navy, the turning point came in 1967, and the first class of ship to reap the full benefit of the change was the Type 42 *Sheffield* class destroyer. These destroyers were the first new class which the Navy had embarked on for ten years. The gap had been largely filled with a long line of the very successful *Leander* class steam frigates, and the time had been occupied with a lot of discussion on future requirements, a number of new ship design-studies — and a Defence Review in 1966 which virtually decapitated the Navy, albeit only temporarily, and negated much of the planning already carried out.

In the period preceding the Type 42 destroyers, two designs are of interest from a machinery point of view. The first was the Type 82.

It is a matter of opinion what designation should be given to this ship. It used to be referred to as a guided missile destroyer (or DLG) — although, of course, the word 'destroyer', derived from the torpedo-boat destroyer of pre-1914 days, is nowadays a complete misnomer. There is, perhaps, a danger in accepting old words which have lost their meaning, and failing to re-define our terms. Certainly it might be argued that it was in part the perpetuation of the name 'destroyer' which resulted in the Navy entering the last war with 80 fleet destroyers bristling with torpedoes — but with almost nothing to use them against except the *Bismarck* (who received her *coup de grace* from a cruiser) — and armed against the might of the *Luftwaffe* with 0.5in machine guns and a 1918 hand-operated 3in AA gun.

In fact, the main task for the wartime destroyer was anti-submarine work, but the 'sloops' of those days, with greater autonomy and endurance, and good ASW equipment (but no torpedoes) were much better suited to the anti-submarine role. Then, in 1943, after a gap of 54 years, we re-introduced the honourable name 'frigate', which, on historical analogy, was a much better designation for the ASW and general-purpose 2000-tonner of those days. It also sounds less warlike than 'destroyer', which may in part account for the anomaly of the Type 82, HMS *Bristol*, being originally designated officially as a frigate. Although planned as a 4000-ton ship, she was to grow during the design-phase to a deep displacement of 7000 tons. By contrast, the Iranian *Saam* class, of 1300 tons are quite firmly designated 'destroyers'. The French confuse the issue further by labelling their 4000-ton *Georges Leygues* class 'corvettes'. The only thing the three classes of ship have in common is that all three have twin Olympus main machinery; but this helps to illustrate one problem which the adoption of gas turbine propulsion has brought to the ship designer. In the days of steam, when the power-requirement for a new class of ship had been established, turbines were designed, probably by the lead shipyard, to meet that specific requirement. There was little 'development' involved, and shore testing was a rarity until the introduction of a new 'standard' steam plant, in the 1950s, for the Type 12 frigates.

In the case of the gas turbine, its very qualities of compactness, efficiency, and reliability were only attained by a long and costly development process, largely paid for by the airmen, but still needing a considerable period from the selection of an established aero-progenitor to the production of a proven marine unit. This had two effects. In the first place, it meant that designers could no longer have their machinery made to measure, but had, as it were, to pick an off-the-shelf product. Secondly the time required to develop a new engine now became larger than the time required to design and build a new ship. It was highly important, therefore, that the navies and the engine-makers between them should get their sums right, well in advance, and be able to produce an engine (or a range of engines) suitable for all the various new classes of warship which would be built in the years ahead.

The Olympus proved a very versatile 'building-block', able to meet the needs of a widely disparate variety of ships which extended eventually from the 700-ton Finnish *Turunmaa* to the 20,000-ton British *Invincible;* but, as we shall see later, the Olympus and the Tyne were insufficient to cover all the longer-term requirements.

In the matter of ship-designation, *Jane's Fighting Ships* has now attempted to resolve the confusion by ignoring individual navies' labels and designating all major surface warships according to their displacement:

Over 10,000 tons — cruisers
7000-10,000 tons — light cruisers
3000-7000 tons — destroyers
1100-3000 tons — frigates
500-1100 tons — corvettes

This is to ignore the common usage amongst those who actually run the things. It seems best to stick to the established labels (even when they have been established for no very obvious reason) and add a word of explanation where necessary, while noting that there is no definable difference between 'frigate' and 'destroyer' in current naval vocabularies.

Whatever we call her, HMS *Bristol* started life as a fleet escort, which meant, in effect, a carrier escort. But she grew in the design-process, and her building cost rose a good deal faster than her displacement. Moreover, her original purpose in life was already disappearing before she was completed, for the political axe had fallen on the projected carrier *CVA.01* in the year in which *Bristol* was ordered.

Nevertheless her construction was continued, although she was to be the only one of her class. She would take to sea the new Sea Dart guided missile, and while she would have COSAG machinery, with steam turbines and main reduction gearing very similar to those already in service in the 'County' class, she would use two Olympus TM1A engines, rated at 22,300hp in place of the four G.6s — the first of the new generation of lightweight gas turbines to be accepted by the RN for a class of operational ships.

The other design in the interim period which is of some interest is the Type 19 frigate. In 1964, their natural lust for power perhaps inflamed by the claims of the gas turbine manufacturers, the Ministry of Defence embarked on a study of a frigate of about 2500 tons, one version of which was powered by four Olympus and four cruising diesels, in a CODOG arrangement. Allowing for installation and transmission losses, this machinery would have delivered over 80,000hp at the propeller, which should have given a dazzling performance in a 2500-ton hull, but the project was dropped — one of a number of still-born designs which any major navy studies in the course of evolving Staff Requirements and developing new classes of ship. The original Olympus TM1, however, as designed for the Type 82 destroyer and used in the Yarrow frigates, had had to be of minimum length to meet the constraints of the Type 82 engine-room, and partly as a result of this it had been an unusually wide exhaust volute. For the Type 19 machinery layout, which called for the installation of four Olympus engines abreast in a hull of some 40 feet beam, a redesign of the exhaust system and the power-turbine mounting was required.

The TM1, with an overall width of 10½ feet, had already been modified, and width reduced to 9ft, for the Vosper Iranian frigates. Now a further streamlining was carried out; weight was reduced, though the full shock-resistance was preserved, overall width was brought down to 8 feet, and performance was, in fact, slightly improved, to produce the standard design of exhaust system and mounting arrangement which is still in service today.

At the same time that the rear end of the Olympus was being modified, improvements were also being made to the front end. This was the start of a process which we have already observed in aero-gas turbines — a process of continuing development and improvement, which can benefit both new engines and those already in service by introducing the necessary modifications into them when they are next removed for routine overhaul.

The changes now made to the Olympus, which affected only the internals of the gas generator, introduced a modest degree of blade-cooling — a technique widely used in aero-engines. At the entry to the turbine, the original Olympus, at its maximum power of 24,000hp, reached a temperature of 1155°K, which was a safe limit for marine operation. Further downstream in the turbine, temperatures were, of course, lower, and could be raised, to produce higher power, without overstepping the safe limit.

The Navy needed some increase in Olympus output to meet the full-power requirement of its new generation of ships, and it could be calculated that the Olympus could be raised to 28,000hp without risk, provided that the first row of stator blades in the turbine — the hottest point — could be cooled. To obtain this power, a turbine entry temperature of 1220°K was required, but with gases from the combustion chamber at this temperature, quite simple stator blade-cooling could keep the metal temp-

erature down to the same level as before.

This process has been taken much further in aero-engines; it was later to be extended again in the case of both the Olympus and the Tyne, and it is also used in the marine versions of later engines. In a marine engine, however, particularly one which has to stand up to the rapid manoeuvring, the shock and blast effects, and the contamination of intake air associated with warship operations, there is a common-sense limit to the process, which has been extended gradually with experience, and with the introduction of improved high-temperature blade materials.

While the Olympus was being streamlined and uprated, the Tyne was also being prepared for its debut as a marine engine, by the now established process of 'marinisation' and by extensive shore-testing, both at Rolls-Royce's engine works and the National Gas Turbine Establishment. The basic alterations in the engine layout have already been touched on in the last chapter. The marinisation included increasing of bearing capacities, strengthening of casings to withstand shock, changes in material to resist corrosion, and some inevitable development in the combustion chambers to reach the objectives of burning diesel fuel with a clean exhaust, and obtaining long 'life'. The external changes, and the design of the mounting system, followed very closely the pattern of the Olympus. The engine was mounted on a short heavy base, with the unit as a whole designed to withstand high shock-loads (in the case of the Tyne, the figure was 40G). The turbine was surrounded by an enclosure which provided sound-and heat-insulation, damage-protection, and isolation of air contaminated by nuclear fallout. The air intake was fitted with a cascaded

*The **Amazon** and her sisters, the first COGOG frigates built for the Navy, proved very popular with the Fleet.*

right-angled elbow piece, to straighten out the incoming air-flow, and the whole engine, with its own local controls, lubrication system, fire-extinguishers, fuel pumps and filters, was produced as a self-contained 'module'. This could be used without any alteration, other than 'handing' the direction of rotation of the output shaft and the siting of controls, in a variety of different ships, so that as the new classes came into service, the training of ships' companies in operation and maintenance, as well as spare parts supply, would increasingly become common throughout the Fleet.

The policy of standardising equipment between different classes of ship was not confined to the gas turbines, but was being pursued in all areas of new ship-design. As far as propulsion machinery was concerned, it was possible to produce almost a complete standard package, comprising not only the two Olympus and two Tyne modules, but the main reduction gearing, uptakes and downtakes, shafting, propellers and controls — the Main Propulsion Machinery Package, or MPMP. The design of this was developed for the Type 42 destroyers, and it was repeated, with minor changes, for the next two classes of ship on which the Royal Navy embarked, the Type 21 and Type 22 frigates. By the time that the design of the Type 42 was completed it had been accepted that the new generation of ships should abandon the reversing gears of the 'Tribal' and 'County' classes, and use controllable pitch propellers for manoeuvring, so the main reduction gearing could now be greatly simplified. The Olympus-Tyne combination had one disparity, in the output shaft speed of the power turbines, for in both cases, these were designed to run at the optimum speed from the gas turbine manufacturer's point of view. This resulted in the Olympus, at 28,000hp, having an output shaft speed of 5660 rpm, and the Tyne, at 4250hp, with a much smaller power turbine, running at 12,750 rpm. To match the power-speed characteristics of the ship's propeller (where the power required varies roughly as the cube of the speed), the Tyne output shaft speed corresponding to the Olympus's 5660 rpm was 3300 rpm. A small primary gear box was therefore built into the standard Tyne module. This got rid of the awkward mechanical problem of having a high-speed shaft, at nearly 13,000 rpm, connecting the engine to the quite separate main reduction gearing, and it also, by using a suitable primary gear ratio, enabled the main reduction gearing to be designed with a single input pinion, the Olympus and Tyne driving the same shaft from opposite ends of the gear box, through separate clutches.

The Type 42 was intended to be the Navy's main guided missile destroyer for a long period, and a class of some 20 ships was originally envisaged. The armament included the new 'Sea Dart' SAM, a single 4.5in gun, and Mark 46 anti-submarine torpedoes; she was fitted with medium range sonar and carried a 'Lynx' helicopter — a ship providing area anti-aircraft defence, with an added surface-to-surface capability in the 'Sea Dart', and good anti-submarine escort capability.

The Type 42 was a Ministry of Defence design, and the first class, HMS *Sheffield,* was ordered in November 1968, but before this the Ministry of Defence had taken the unusual step — for the first time in 20 years — of placing a design contract with a commercial shipyard. This was for a much smaller frigate, of 2500 tons, and the contract was placed with Vosper-Thornycroft and Yarrow in February 1968. The designers' task was a demanding one, for the staff requirement called for full frigate capability, including adequate anti-aircraft point defence, effective task force escort defence against surface or submarine attack, and the ability to carry a 'Lynx' helicopter, all within the 2500 ton displacement limit, with a 4000-mile endurance. The design had to maintain RN standards of stability, sea-keeping, damage control, and world-wide operational ability. Armament was necessarily more restricted than in the Type 42, and 'Sea Cat' SAM was carried in place of 'Sea Dart', though later ships mounted four 'Exocet' SSM, and all had the same anti-submarine capability as the Type 42. The propulsion plant was modelled on the Type 42, with the same Olympus and Tyne modules and, allowing for minor changes due to the reduction in ship size, the same standard Main Propulsion Machinery Package. The result was the Type 21 frigate, of which the first of class, HMS *Amazon,* was ordered from Vosper in March 1969. A smaller and simpler ship than the Type 42, *Amazon* commissioned in 1974, nine months ahead of *Sheffield,* and was thus the first ship to take the

*H.M.S. **Sheffield** and H.M.S. **Amazon**, first of class of the Type 42 destroyers and Type 21 frigates.*

*"... the nearest... to the NATO standard frigate..." . HNIMS **Kortenaer**, first of twelve new Dutch frigates. Two are also building for the Greek Navy.*

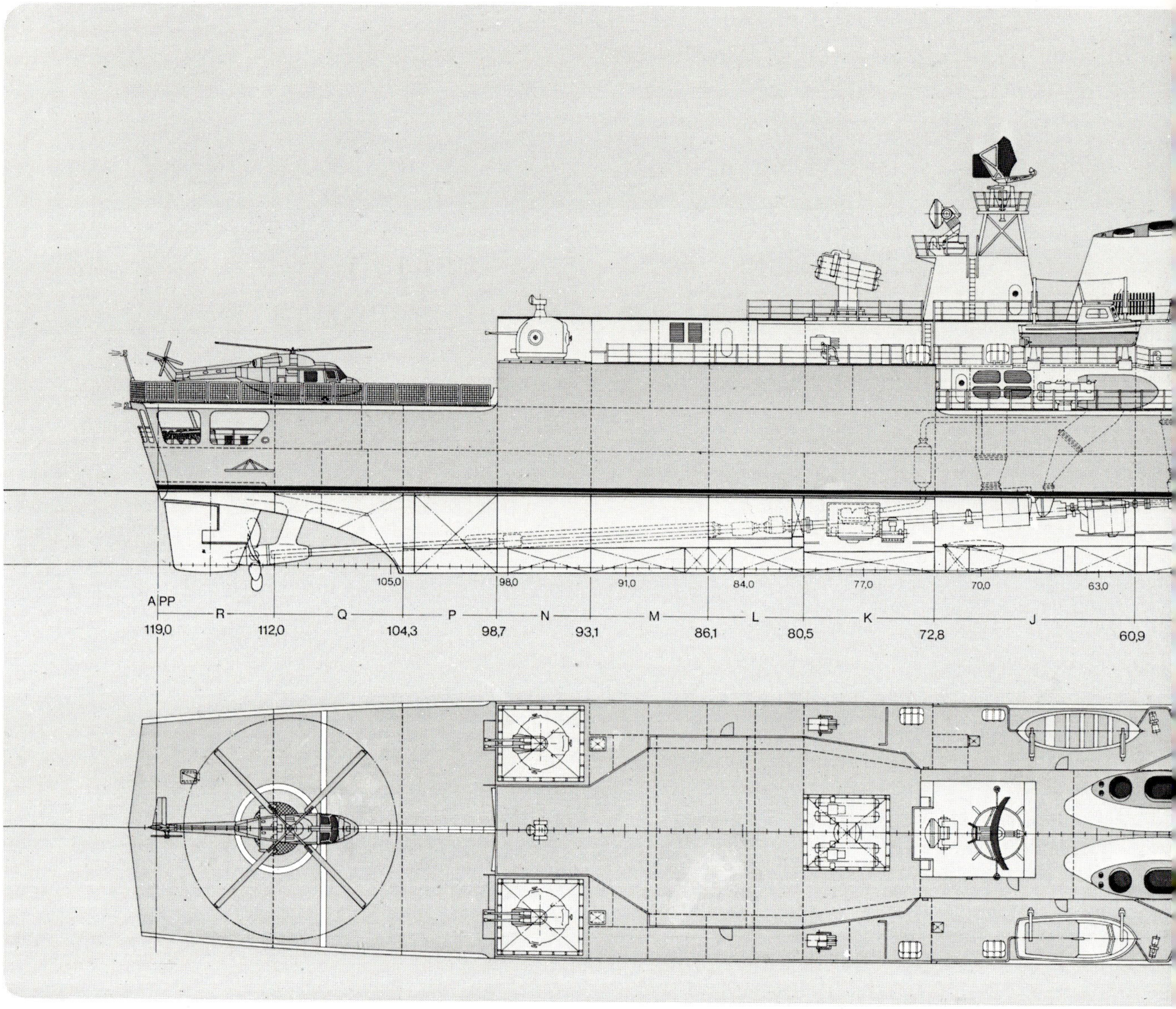

The Blöhm and Voss MEKO 360 is the largest of their standard series of frigate and corvette designs, and a number of options are available both for weapons and sensors, and for machinery. The first ship, built for the Nigerian Navy (**RIGHT**) uses Olympus CODOG machinery; four similar frigates for Argentina will be powered by the standard Olympus-Tyne COGOG package.

H.M.S. **Battleaxe,** 4000 ton Type 22, the latest and most sophisticated class of British long-range anti-submarine frigate.

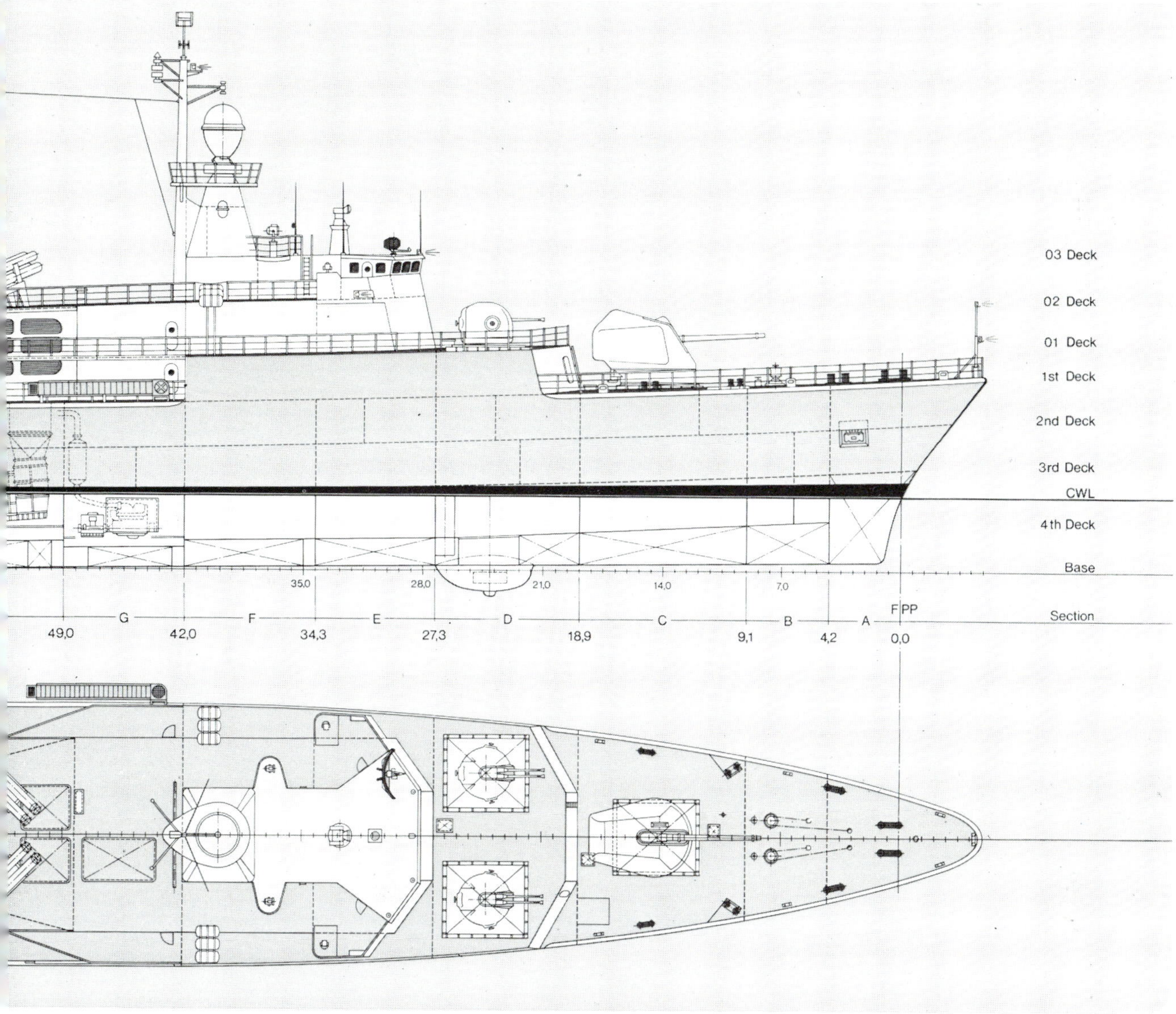

03 Deck
02 Deck
01 Deck
1st Deck
2nd Deck
3rd Deck
CWL
4th Deck
Base
Section

35,0   28,0   21,0   14,0   7,0

G   F   E   D   C   B   A   F PP

49,0   42,0   34,3   27,3   18,9   9,1   4,2   0,0

new standard COGOG machinery to sea.

In both the Type 42 and the Type 21, a number of new features were introduced, apart from the gas turbine machinery, and a lot of study was given to the problems of maintenance, and to the development of the philosophy of 'repair by replacement'. It is interesting to see the effect that these developments, in particular the introduction of the new engines, had on manning. It is possible to make a reasonably fair comparison between the Type 21 and the *Leander* class steam frigates, for both had a broadly similar displacement and capability. The *Leanders,* developing 30,000hp, required an engine room complement (including electrical personnel) of 2 officers and 54 men; in the Type 21, with 60 per cent more power, the engine room complement was reduced to 1 officer and 41 men. This drop in technical manpower of 25 per cent was, of course, due in part to the cutting down of auxiliary onboard maintenance by increased 'repair by replacement', but the greater part of the reduction was the result of installing all-gas turbine machinery.

The Type 21 frigates have come in for some criticism in recent times, but the general verdict of those who sailed in them has been very good. Most people have agreed that the design was well thought out, and made the best use of the restricted space available. No doubt, with hindsight, it can be seen that a slightly larger hull — which adds relatively little to the cost, and will not necessarily have any adverse effect on performance — would have been an advantage, and would have made it easier to fit improved weapon systems when the need arose later in the ships' life. The same can be said of many other ships. The *Leander* hull was broadened for later ships of the class, and is now being stretched 40ft (in India); later Type 42 destroyers are also being stretched 40ft and broadened, and even the new Type 22 frigates are being stretched 40ft. It is an unfortunate fact that ship-designers are commonly in the uncomfortable position of being squeezed between the Naval Staff and the Treasury. Perhaps it is no coincidence that the proverb condemning false economy uses a nautical image.

The Type 22 or *Broadsword* class frigate, mentioned above, was the latest of the new family of British frigates and destroyers to come into service in the 1970s. Built primarily for long-range ASW operation and the high-endurance task of 'policing' the so-called Greenland-Iceland-UK Gap, these 'frigates' displace over 4000 tons. With the 'Exocet' SSM and the new 'Sea Wolf' SAM, the 4.5in gun could be sacrificed in favour of more sophisticated sonar and electronic equipment, and there was a more generous allocation of space within the ship than had been the case in previous classes.

Again, the standard Olympus-Tyne COGOG machinery was used. This adoption of a standard propulsion package led to considerable economies in provision of spare parts and crew-training, the more so as the same arrangement, and the same standard engine modules, were also being adopted by other navies, so that sharing of logistic support became possible.

Such pooling of resources was initiated in 1971, when the Royal Netherlands Navy embarked on a major new building programme to equip the fleet with modern guided missile destroyers and frigates, to replace the ageing *Holland* and *Friesland* classes. This programme comprised two destroyers, *Tromp* and *De Ruyter,* of 5400 tons displacement, and a class of 12 frigates of 3800 tons, both classes of ship using the same Olympus-Tyne machinery arrangement as the British ships, though with Dutch gearing, propellers, and controls.

By common consent these Dutch destroyers and frigates have been acclaimed as excellent ships, and the *Kortenaer* class frigates are probably the nearest we shall get to meeting the requirements of that long-sought mythical creature, the NATO Standard Frigate. In addition to their technical excellence in design, they represent a good spread of NATO's particular talents, with Dutch radar, controls, and electronics, Canadian sonar, American ASW torpedoes and 'Harpoon' SSMs, NATO 'Sea Sparrow' SAM, Italian OTO-Melara gun, the British 'Lynx' helicopter, and British engines.

They are already being built for one other NATO navy, the Greek, and the hull-form has been used for the new German *Bremen* class frigates, although, for a variety of reasons, some equipment, including engines, has been changed in these last, and American gas turbines are installed with German diesels in a CODOG arrangement. Germany, the home of the diesel engine, has always been pre-eminent in this field, so the choice of diesel-cruise in this case was understandable, and it has been left to Blöhm and Voss, carrying on their great warship-building tradition, to introduce Olympus-Tyne powered ships for more distant navies.

One of these was the Argentine Navy, for they were the first overseas navy to adopt the standard COGOG propulsion arrangement, even before the Dutch. In 1970, they ordered two Type 42 destroyers, built to almost exactly the same specification as the British ships, the first, *Hercules,* by the lead-yard for the whole class, Vickers, and the second, *Santisima Trinidad,* in the Navy's own yard at Rio Santiago. The Navy had a further requirement for frigates, but it was not until 1978 that an order was placed with Blöhm and Voss for four of their MEKO 360 type frigates. This design, of 3600 tons, is the largest of the MEKO 'family' of frigates and corvettes, which features the Blöhm and Voss philosophy of providing a complete range of weapons and sensors in interchangeable containers or mounting-units. This method of packaging equipment, although it may entail some sacrifice in weight and space, has the advantage of cutting down building time — and therefore cost — as well as allowing a wide choice of weapon systems, both initially, and later on, by conversion in service.

The Royal Belgian Navy's **Westdiep** *uses a single Olympus with a split drive, and twin Cockerill diesels. The design and building of these large corvettes was a major step forward for the Belgians.*

BELOW: *George Leygues, first of the French C.70 corvettes, of 4000 tons. Twin Olympus are combined with Pielstick diesels in a CODOG arrangement.*

One MEKO 360 had already been ordered from Blöhm and Voss by the Nigerian Navy in 1978, but this was a CODOG version, with two Olympus and two MTU diesels. The Argentines, with the Tyne already in service, and mindful of the need to economize in maintenance, chose the all-gas turbine version with the standard Olympus-Tyne package.

Returning to the European scene, the Dutch decision to use the same machinery as the RN enabled the two navies to set up a pooling arrangement for both spare parts and spare engine units, as well as to benefit from each other's operating experience. In so far as the Olympus was concerned, this collaboration was soon extended to two other navies, to form a 'European Club'. The first was the Belgian Navy, which in 1973 had embarked for the first time on a major warship programme, designing and building four well-armed small frigates of the *Wielingen* class, which we referred to in an earlier chapter.

The next members of the European Club were the French, who laid down the first of their C70 *George Leygues* class at Brest in 1974. With a wealth of well-developed weapons and machinery to choose from , there was little need for the French to look beyond their own frontiers to arm and equip these 4000-ton so-called 'corvettes'. Weapons and sensors, all of French origin, including 'Exocet' SSMs and 'Crotale' SAMs, two Westland-Aerospatiale 'Lynx' helicopters were carried, and the Pielstick diesel was used as a cruising engine. For high-speed operation, however, no French gas turbine of appropriate size was available. The Navy had already experimented with gas turbine-propulsion in the *Commandant Riviere* class frigate *Balny,* which commissioned in 1970, powered by two Pielstick diesels and a single SNECMA M38 gas turbine in a CODAG arrangement. The M38, however, developed from the 'Atar' jet en-gine, provided only 14,000hp. For the *George Leygues* class, two Olympus were therefore used in a CODOG arrangement.

On the other side of the world, the Japanese Maritime Self-Defence Force had started planning to introduce gas turbine propulsion in 1970 or earlier, but it was not until 1977 that the first ships in the new programme were approved. Two new classes were planned initially: a small 'destroyer-escort' of some 1500 tons, and a larger fleet destroyer in the 3000-ton class, and the Defence Agency carried out a highly detailed review of all aspects of the new ships, including a long investigation of alternative machinery arrangements. Gas turbines were favoured because of the economy in manpower and the increased ship-availability which they afforded, and from the equipment available in 1977 the final choice was a CODOG arrangement for the destroyer escort, with a single

*The Japanese **Ishikari**, first of the new destroyer escorts for the Maritime Self-Defence Force, and the beginning of a major re-building programme. Present ships of the class are powered by a single Olympus with a Mitsubishi cruising diesel.*

Olympus and a single Mitsubishi diesel, and Olympus-Tyne COGOG in the destroyers.

The success of the all-gas turbine propulsion package virtually eliminated steam plant from further consideration, although diesels would continue to play a large part in warship-propulsion and would gain in popularity in the mid-1970s, with the advent of oil price-increases and the spectre of world oil-scarcity. But the gas turbine's success also brought problems, for it was being pressed into service in an enormous range of ships. The Olympus-Tyne combination was being used in hulls ranging from 2500 tons to nearly 5500 tons, and while the Olympus, uprated to 28,000hp, provided sufficient top speed in all the existing ships, there was a growing need for increased cruising power in the larger destroyers.

Some improvements could be made without difficulty, since the Tyne was by no means stretched to its limit, and had a margin for a 25 per cent increase in power by using cooled turbine blades, without going beyond safe limits and without any major re-design. Work on this uprated version of the engine was started in 1972; a cooled rotor blade was produced, and an improved blade-material, recently developed by the metallurgists, was selected; but it was not until 1978, after extensive shore-testing, that the new standard of engine was introduced into service, in HMS *Ambuscade* and in the first of the Dutch frigates, HNIMS *Kortenaer.* This provided a considerable improvement in cruising speed, for not only was the power increased by just over 25 per cent, to 5340hp, but the increase could be sustained at higher ambient air-temperature. So, in tropical conditions, where the original Tyne had been reduced to 3660hp, the uprated version still delivered 5100hp — an increase of almost 40 per cent, and sufficient, in a typical frigate, to put an extra two knots on the cruising speed. As the new engine was interchangeable with the old, and as the uprating had been anticipated in the original design of the Tyne's primary reduction gear, it was possible to convert existing engines to the higher rating as they came in for periodic overhaul. Thus the whole Fleet could take advantage of the improvement.

A similar type of blade-cooling, covering the first row of rotor blades and the first and second rows of stator blades, had been introduced in the industrial version of the Olympus engine in 1974. A standard of Olympus was therefore available for marine applications, which could produce 29,600hp if required — an increase of some 25 per cent under tropical conditions.

Another problem which had exercised the minds of the Royal Navy from a very early stage in the development of gas turbines was the difficulty of providing astern-power. Reversing in the gearbox, such as had been used in the 'Tribal' and 'County' class ships, introduced considerable complication, and controllable-pitch propellers were at first viewed with some misgiving, from the point of view of reliability, and efficiency, and underwater noise. Since then, CP propeller design has progressed, and they have been accepted for nearly all new ships; at the same time, a new and simplified system of reversing in the main reduction-gearing has been developed by Franco-Tosi in Italy. In 1967, however, the problem still appeared to be sufficiently important to merit examination of other possible solutions, and an ingenious and laudable attempt was made by Rolls-Royce to produce a Reversing Olympus.

This consisted of a standard gas-generator coupled to a centripetal, or inward-flow, turbine. This was fitted with swivelling nozzles which could be rotated to give ahead or astern power to the impeller. Such an arrangement not only solved the problem of reversing, but it also saved space, because the Olympus gas-generator, having its axis at right-angles to the reversing turbine shaft-line, could conveniently be mounted vertically in its own air intake trunk, and would therefore, in effect, occupy no space at all. With the gas generator disposed of, the centripetal turbine needed very little fore-and-aft space, so it could be accommodated in a very short engine room. It all sounded marvellous; but it had its snags.

There was some slight reduction in efficiency in ahead drive, though rather less than anticipated. When running astern power was much more limited, since while, the nozzles were driving the rotor one way, the exducer section of the impeller was still trying in vain to drive it the other way; but the efficiency was still 60 per cent and power was more than adequate for manoeuvring. A bigger problem was the size of the inlet volute and the impeller. In practice, forging limitations restricted the size of the impeller and limited the maximum power attainable to about 10,000hp. Although this restriction could be overcome by splitting the gas-generator output between two turbines, each turbine volute would measure 11ft in diameter, and the complete unit would be much heavier and more massive than the equivalent conventional axial-flow power turbine.

Rolls-Royce built a third-scale model and used a cold-air supply to blow it through its paces. It worked very well; but the Navy, perhaps wisely, decided to devote its resources to improving the CP propeller.

To date, there have only been two exceptions to this rule. The first was in HMS *Invincible.* It will be recalled that the politicians had abolished aircraft carriers in 1966, but although the tree had been officially cut down, new shoots soon started to grow from the stump. By 1969 plans had been developed for a large helicopter-carrying cruiser, known officially as the Cruiser Assault Helicopter or CAH, but usually referred to, for some mysterious reason, as the Through-Deck Cruiser. The word 'aircraft-carrier' was never mentioned. Indeed, in order to disguise the fact that the ship would have a flat top, it was even suggested that she might have an island on each side, with the flight deck discreetly hidden in the middle. Whatever was to happen on deck, it was clear that a new machinery layout would be needed down below, as the preliminary design gave a ship of about 19,000 tons.

By the end of 1969, it had been decided that propulsion would be by four Olympus engines driving twin shafts. To transmit over 50,000hp through a CP propeller was a considerable step, and it was therefore decided to use a conven-

The first Olympus for the new Japanese programme on test at the Kobe works of Kawasaki Heavy Industries. The Japanese Defence Agency selected combinations of the Olympus, Tyne, and Spey to provided the optimum machinery arrangement for three new classes of ship, ranging from 1500 to over 5000 tons.

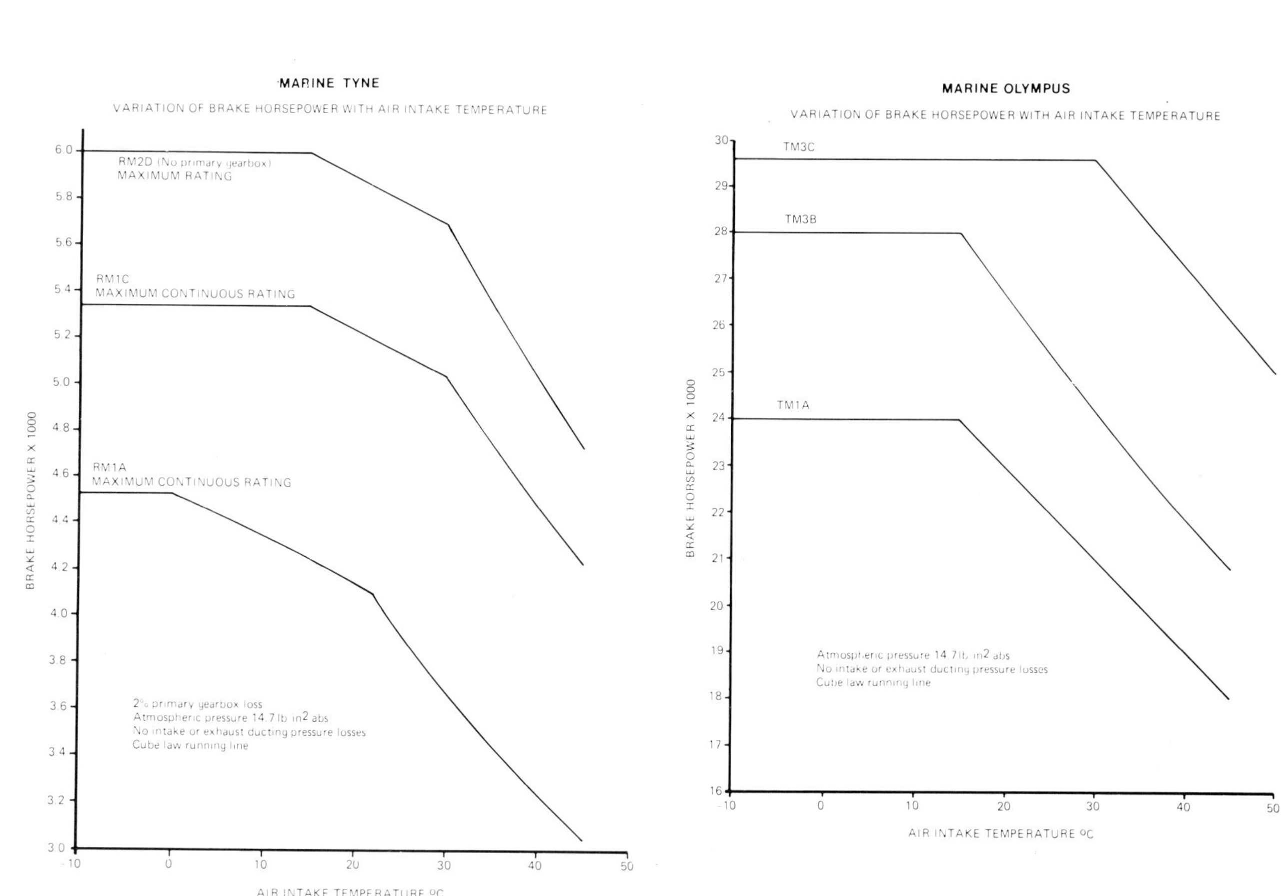

Over the years, both the Olympus and the Tyne have been uprated by some 25% — or 40% under tropical conditions — through the development of improved turbine blade material and blade cooling techniques, but there is a limit to this process, particularly for marine engines.

tional fixed-pitch propeller and a reversing gearbox. This introduced a major piece of un-tried equipment into the machinery; in addition (since it had been decided to have a single island, and to display the flight deck in all its nakedness, at least from the port side), it was necessary to duct the port engine exhausts across the full width of the ship under the hanger deck, and introduce right-angled bends at each end. To test the ship-installation, and, in particular, to check the operation of the reverse-reduction gearing under extreme manoeuvring conditions, the complete port shaft set of machinery was built in a shore test-house by Rolls-Royce. This comprised the two Olympus, the reversing gearbox, intakes running from either side of the ship below the flight deck, exhausts ducted across to the position which the after funnel would occupy, a low-speed dynamometer on the propeller shaft, and complete ship-controls.

This type of shore-testing has been done elsewhere — with gas turbine plant in France, Holland and the USA, and with steam plant in a number of instances — but never with quite such a complete replica of a whole ship-installation. It was to prove invaluable, for some modifications to the equipment did prove necessary, but these could be made without difficulty during the shore testing period, whereas they would have involved major surgery and many months' delay had they been required after trials had started in the ship itself.

As it was, the sea trials proved extremely trouble-free, and the *Invincible* class have so far been shown to be excellent ships, even though the sweet smell of success may be somewhat marred by an ominous odour of mothballs.

The second use of a reversing gearbox, still only a paper exercise, was planned for the Royal New Zealand Navy. This was an extremely interesting application since it involved the conversion of an existing 30,000hp twin-shaft steam plant to a simple all-gas tur-

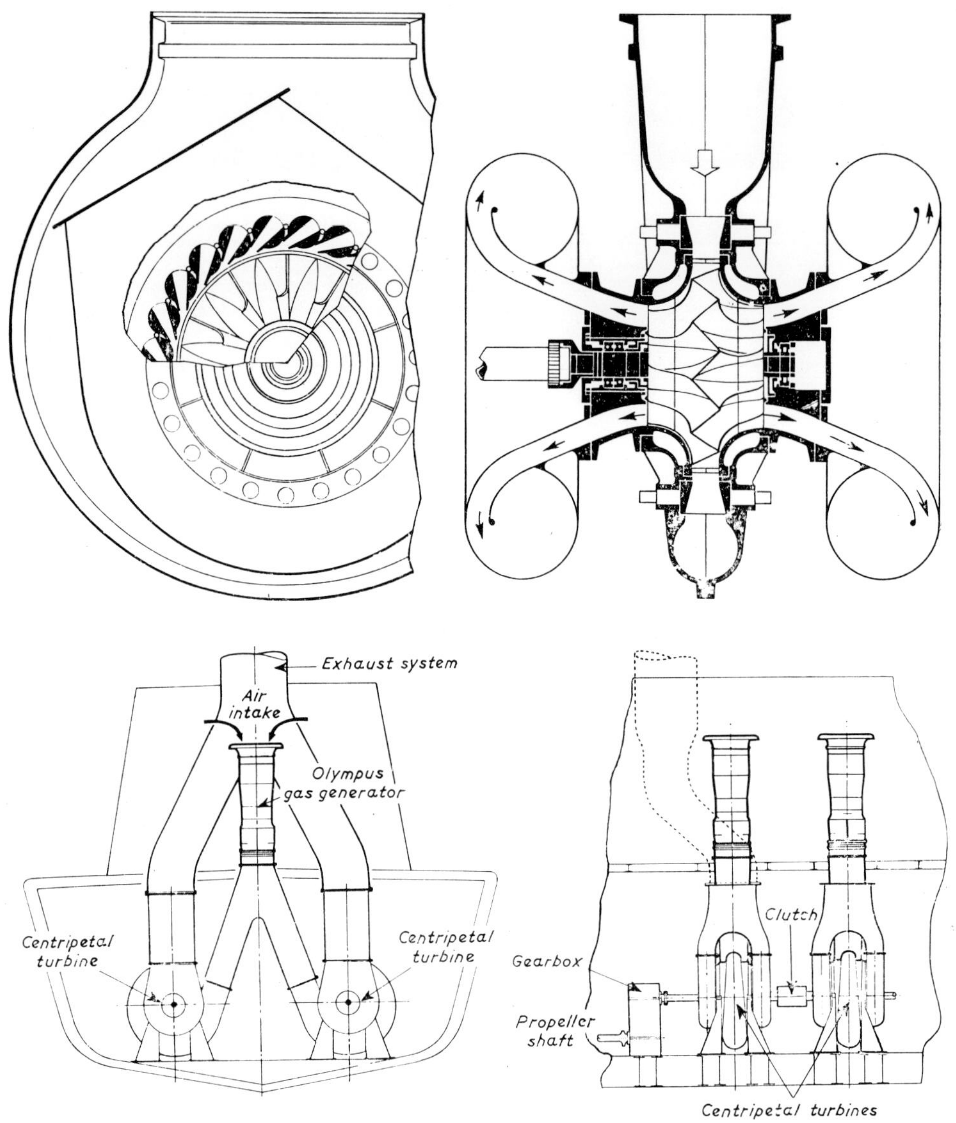

The centripetal turbine. A laudable but still-borne attempt to produce a reversing Olympus. Reversing gears or controllable-pitch propellers still provide the best solution to the problem of manoeuvring.

bine installation, with a single engine on each shaft in a COGAG arrangement.

The New Zealand Navy had a need for a Resource Protection or offshore patrol vessel which would also be used for training. Her duties would entail spending long periods away from base and, from a machinery point of view, the main requirements were:

1. High availability — and therefore low maintenance and good reliability
2. High endurance
3. Low operating costs — in manpower, upkeep and fuel
4. Medium top speed

The Type 12 frigate *Taranaki* was earmarked,

but though she had a good hull — the Type 12 and *Leander* hulls have stood the test of time extremely well, in terms of both sea-keeping and good lasting properties — she was an old ship, built in 1961, expensive in fuel and manpower, with machinery increasingly difficult to maintain, and insufficient endurance for her new role.

The Navy decided, after examining all the alternatives, to scrap the *Taranaki* steam machinery and replace it with two Tyne engines. These would drive through the existing main reduction-gearing, modified to include the Franco-Tosi reversing gear and SSS clutches, which would make it possible to use the existing shafting and propellers. The gas turbines could be

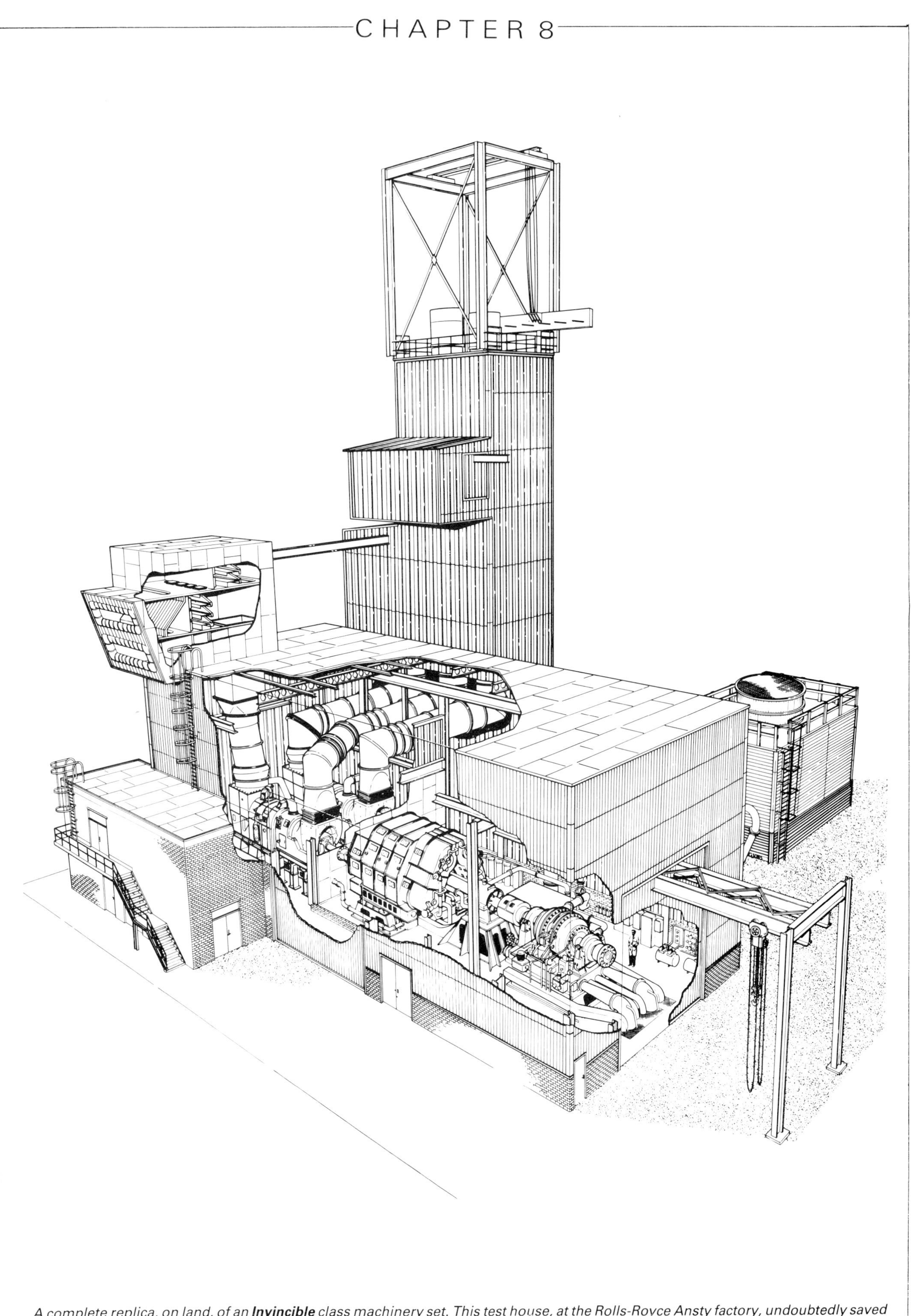

*A complete replica, on land, of an **Invincible** class machinery set. This test house, at the Rolls-Royce Ansty factory, undoubtedly saved much time and money, by eliminating troubles before **Invincible** went to sea.*

accommodated within the engine-room in place of the steam turbines, and the boiler-room would be available for new diesel generators, a Ship Control Centre, and additional fuel stowage. This new machinery arrangement reduced maximum speed to a little over 22 knots, but that was sufficient for the new role, and the re-engining had a number of quite dramatic advantages.

It cut fuel consumption by 40 per cent. With the extra fuel-stowage available, this meant that endurance was increased to 2½ times **the** original figure. It reduced manpower; it increased availability; and it gave a new lease of life to an ageing ship, at an economic cost.

The plan has been dropped, but it raises interesting possibilities. One can envisage the same type of conversion being carried out, with similar benefits, in a number of existing Type 12 and *Leander* class frigates, including the New Zealanders' own *Canterbury* class, as well as in the numerous other steam frigates in the 2000-3000 ton/30,000hp category. In most cases, however, there would be a requirement to maintain the present maximum speed of about 30 knots. The need would be for a high-efficiency engine in the same power-bracket as the existing steam plant, or some 15,000hp.

A unit of this power, filling the gap between the current 5000hp cruise and 30,000hp boost gas turbines, could obviously have other uses; but until a short time ago, no such engine has been available in the world's range of marine gas turbines.

*H.M.S. **Invincible**, first of the new British aircraft carriers, of 20,000 tons, powered by four Olympus.*

A NEW GENERATION

It is fashionable in some aero and marine circles to talk of 'second-generation' engines, as though gas turbines can all be neatly classified in two well-defined groups. This label is of more value as an advertising slogan than as a description of any particular unit in the long development story of the gas turbine family.

It is, of course, possible to trace a line of improvement over the years, both in the efficiency of blading and other individual components, and in the working limits of temperature and pressure which are the main determinants of specific power and efficiency. In some cases, these improvements are applied to a later engine in the same family, in some cases to a new design. It is possible, also, to distinguish a 'first generation' of marine gas turbines in the sense that the requirements of ship-design and production make it necessary, in introducing any new piece of major equipment, to call a halt at some point to the otherwise endless process of refinement and modification, and say 'this is now a workable machine which meets our requirements, and we will establish it as a standard unit'.

That point was reached by the RN in 1967. For the next ten years, it was not merely a question of sitting back and letting the factories turn out standard engines. The business of development and improvement continued, but it was essential that a major part of this should be directed to refining the existing engines without deviating from basic designs, so that a whole range of ships, which would be built over a span of 10 years and remain in service for a further 20, could all benefit from the work. In the case of ship propulsion machinery, this implies starting with a unit which is the best that can be produced at that time, but which still has potential for further 'stretching', to meet changing needs and (nearly always) a growth in the size of the ships.

We have seen how the Olympus and the Tyne, as well as the earlier Proteus, met this requirement, with increases in power and, through smaller and less spectacular changes, a gradual improvement in reliability and the length of running-time between overhauls.

There are two limitations on this process. In the first place, even if an individual engine type can be 'stretched' a long way — *vide* the case of the aero-Olympus — there comes a point when the design has departed so far from the original that it has lost all commonality, and all the advantages of service experience and standardisation. There is no object in going beyond this stage, and further requirements are better met by picking a new basic design as a fresh point of departure.

The second limitation is on improvements in efficiency. The normal line of development of a given design of gas turbine is via increases in compression-ratio (through the addition of further stages to the compressor) and increases in turbine temperature (made possible by improvements in turbine blading). Increasing turbine temperature, by burning more fuel, increases the maximum power, and we have seen this method used in both the Olympus and the Tyne. This will usually produce some slight increase in efficiency at the maximum power, but the overall fuel-consumption is not improved at all. In fact, if the uprating is achieved by increased blade cooling, then, if both are running at the same power, the uprated engine will be less efficient than its predecessor, because some of the air from the compressor is being diverted to cooling the turbine blades.

Increasing pressure-ratio, accompanied by

some raising of the turbine temperature, can have a marked effect on efficiency, but there is a limit to this process, for if it is carried beyond a certain point, it involves re-design of major portions of the engine, and not merely the addition of another row of blades at the front of the compressor. Along this road, we will soon meet the point where it is better to start afresh with a new basic design. This is partly because the design of compressor-blading itself has improved over the years, and a more recent design of compressor will therefore produce the same compression-ratio with a smaller number of stages and fewer blades, at less cost, and with greater efficiency.

We can see this happening in the case of the Avon and Spey engines, which are both within the same power-bracket. The Avon, starting in service as a military jet engine in 1951, has had a long family history, and current versions of the civil aircraft engine (which is the type from which the industrial Avon was derived) have had two additional stages added to the compressor, which now has a compression ratio of 12:1, produced by 17 stages of blading. The more modern Spey, in its latest form, has a much higher compression ratio of 18:1, but with one fewer stages of blading than the Avon.

The successive developments of the marine Olympus and the marine Tyne, which we noted in the last chapter, enabled them to meet all the needs of the new ships and give good service. The process of improvement continued, and basic power-output could still be raised further; but parallel with the refinement of the two standard engines, the Navy was also looking for basic new designs, from which to develop new types of marine engine to meet differing future needs. There were two reasons for this quest, which one might call a theoretical and a practical reason.

The theoretical reason was the fairly obvious one that we had a small cruise engine and a large boost engine, but none of intermediate power. This was appreciated when the Olympus and Tyne were first introduced. In 1968 the Minister of Defence, announcing to the House of Commons the Navy's decision to use gas turbines in all future ships, added to his statement: '....That leaves a gap in power-range between the Olympus at about 25,000hp and the Tyne at about 4500hp, and we are now considering whether we need an intermediate engine and, if so, what it should be'. The 'considering' was to be a long process, involving the investigation of various alternatives, precisely because the Olympus-Tyne combination (with the 'gap' reduced a little by the uprating of the Tyne) fitted the operational requirements of the Fleet very well, and provided good fuel economy. So, although it was appreciated that gas turbines must be considered as more or less fixed 'packages' of power, which could not be redesigned to suit the precise requirements of each new class of ship, and although, in theory, there was a yawning gap between the Olympus and the Tyne, there was no practical requirement to fill it.

The effect of the gap is plain to see, however, if we look at the fuel consumption — and hence the endurance and fuel cost — of any COGOG frigate or destroyer at varying speeds. Up to about 18 knots, economy is good throughout the cruising-speed-range; at high speed, fuel-consumption is of course much higher, because much higher powers are being used, but the engine efficiency is still good. It is better than steam, and although the Olympus is not quite so economical as the Tyne, these high powers are in any case seldom used. The position is not quite so good at intermediate speeds. Whereas our typical frigate can steam 7 miles on a ton of fuel at 18 knots when running on two Tynes, when it changes to a single Olympus, even at the same speed, the endurance falls to 5 miles per ton, and this reduces further as speed is increased. These are only approximate figures, calculated for an imaginery ship of 3000-4000 tons; but they allow for a normal 'hotel' load, for installation and transmission losses, and for additional loss when one shaft is being trailed.

While you are content to cruise at 18 knots (or at whatever is the maximum speed attainable on the cruising engines), this 'knee' in the endurance curve is largely of academic interest. Most of the running is done on the economical cruising engines, and use of high speed is expected, in any case, to be more expensive in fuel; but operation at speeds just above the cruising engine limit is still perfectly feasible, when required.

During the 1970s, however, the pattern operation was changing. This was partly through development in sonar equipment allowing anti-submarine operation at higher speeds, and partly, because a reduction in the number of ships available, while the tasks remained the same, made a more rapid development desirable. Whatever the reasons, the normal cruising and ASW operating speed, hitherto 16-18 knots, was moving up towards the 22-25 knot level. This was quite beyond the range of the Tyne, and if better economy were required at the higher cruising speed, then the only solution would be to use a new and bigger cruising engine, or a new and more efficient boost engine.

As far as the latter was concerned, there was a new and much more efficient aero-engine already to hand, in the Rolls-Royce RB-211, on which full development was started in 1968 for the Lockheed 'Tristar' aircraft. With the engine due to go into service in 1972, in a major new wide-bodied passenger aircraft, there would be a very rapid build-up of flying experience to provide basic background for marine development, and plans were already in hand for an industrial version of the engine.

In 1969, the Ministry of Defence asked Rolls-

THE RB.211 AERO ENGINE

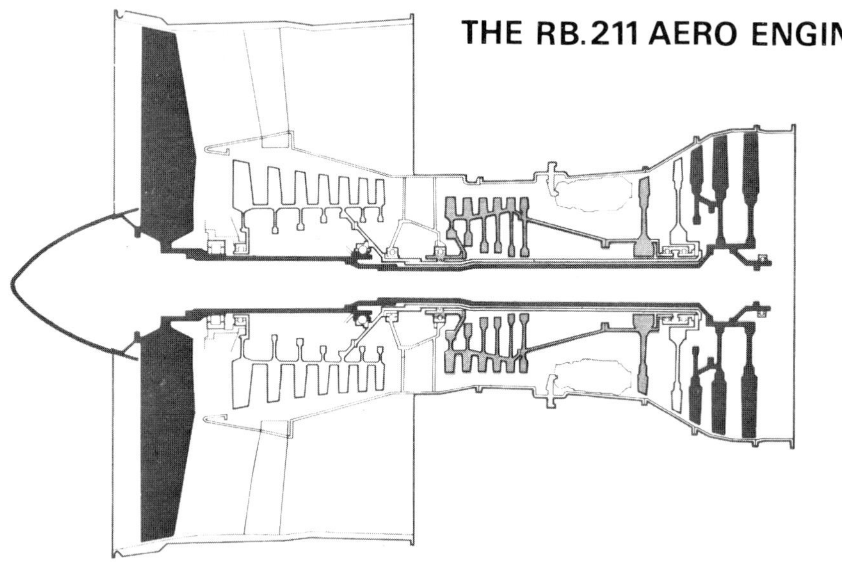

The RB 211 was a three-spool engine, with a large front fan driven by the low-pressure turbine. Conversion for either industrial or marine application involved, basically, removing the fan, fan turbine, and inner shaft, leaving a high-efficiency two-spool gas generator, similar in layout to the Olympus.

For naval applications, a two-stage long-life power turbine, running at about 5000 rpm, was proposed. Power turbines for the industrial RB 211 are of similar design, aero-dynamically. They do not, of course, have to be designed to withstand high shock loads, but this is not needed for merchant ship applications.

THE MARINE RB.211

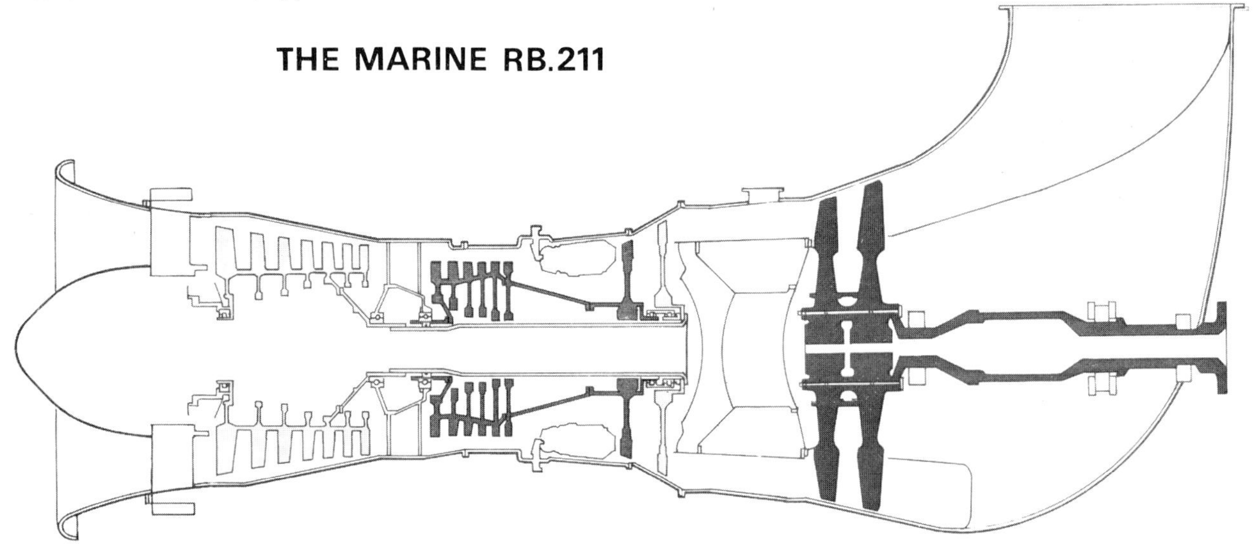

Royce to carry out a Design Study on a marine RB-211. This engine was not quite so easily converted to marine use as the Olympus. In the case of the latter it had been possible to use the aero-jet engine as a gas-generator without any changes to the layout, so that the marinisation was confined to a number of detailed alterations to the bearings, casings, combustion chambers, and controls, and to changes in material. The RB-211 represented a further advance on the two-spool concept used in the Olympus. It had three spools, two of them consisting of the high- and low-pressure sections of the compressor, each with its own turbine, and the third carrying the front fan. By throwing away the fan, together with its driving turbine and interconnecting shaft, and doing some consequential re-design of the front end, the aero-engine could be reduced to a two-spool gas-generator with a configuration similar to that of the Olympus, but with a much higher compression-ratio, and cooled blades designed to run at higher temperatures. A new power turbine of two stages, would be required, as, although the power was about the same as the Olympus, the air-mass flow was much lower and the power turbine inlet-pressure much greater.

This produced a very attractive engine, with a maximum power output of 27,500hp and a specific fuel consumption reduced to the spectacularly low figure of 0.4lbs/hp-hr. Even more important, at half power, or 14,000hp, the fuel consumption was still only 0.47lbs/hp-hr. With such an engine, the endurance of our frigate or destroyer at 18 knots, though not as good as with the Tynes, is up to over 6 miles per ton of fuel.

This was a considerable step in the right direction and the RB-211 would have been a very useful engine for a number of naval applications; but it could not, by itself, replace the existing Olympus-Tyne machinery arrangement, since it could not compete with the Tyne at lower cruising speeds. There was very limited advantage in using it only as a replacement for the Olympus, since, while it would achieve some fuel-saving at higher speeds, this would only be of benefit for a small proportion of the ship's total running time, and the net fuel cost saving would be offset by the higher capital cost of the RB-211.

An engine such as the RB-211 would certainly

Gas turbines have found little favour in normal merchant ship applications, because they cannot compete with the big 'cathedral' diesel, burning heavy residual fuel. They are, however, an attractive type of engine for large natural gas carriers, since these will, in any case, use 'boil-off' from their cargo for part of their propulsion fuel requirement, and this is an admirable gas turbine fuel. Electric drive, with the lightweight gas turbine sited high in the after superstructure, at the foot of the funnel casing, makes a neat layout, with minimum use of space.

have been valuable in larger ships, where it could have been used as the sole means of propulsion in a COGAG installation, such as was used in the *Invincible* class aircraft carriers and in the 7000-ton American guided missile destroyers.

In later years, it was to be used very successfully as an industrial engine, and was to be considered for a number of merchant ship requirements. It could be applied to naval propulsion with little difficulty, if a fresh need for a cruiser or carrier were to arise, but this seems highly unlikely in the Royal Navy, at least within the lifetime of the present Government.

But improving the larger boost engine was not the most efficient way of providing more economical high-speed cruising power in the major part of the fleet — the frigates and destroyers. If a cruising engine were available which had an output sufficient to meet the new cruising speed requirement, with an efficiency to meet the RB-211, then we should achieve a much greater saving in fuel. But to cruise at 22-24 knots would need over 10,000hp per shaft in a twin-screw ship.

It was the need for a larger cruising engine that had prompted British interest in the M.45 aero-engine, which we saw, in an earlier chapter, was being considered in the 1960s as a possible successor to the Proteus. By that time, the British were no longer interested in new engines for small craft, since the Navy's Coastal Forces role had been abandoned, and one reason for the eventual demise of the projected marine M.45 was that, considered as a cruise engine for big ships, it had little power advantage over the already well-developed Tyne. It was possible to envisage stretching the M.45 to 7000, perhaps 8000hp, and the Bristol-Siddeley hybrid, the 'Centaur', was given a maximum rating of 8000hp. But this was not enough; what the Navy really needed was a unit of at least 11,000hp and perhaps more.

Increasing the power of the cruise engine to this level to meet the higher cruising speed requirements, raised other questions. We have seen how the peculiar conditions of naval ship operation — with most of the time spent at relatively low power, but with much higher power still required to be available at need — made the cruise-boost COGOG concept attractive. But this arrangement was fundamentally unsound in one respect, since it meant that at maximum speed, in the heat of battle, chasing — or being chased by — the enemy, half the ship's engines were standing idle. When the disparity between the cruise engine and the boost engine was sufficiently large, this was not much loss. Two Tynes clutched in on top of two Olympus at full power would only add a knot or so to the maximum speed, and this small gain would be offset by the need to complicate the transmission with the introduction of a two-speed gear into the Tyne drive — for the engine would have to run at full power by itself to meet the cruise-requirement of 18 knots, and again at full power with the Olympus at 30 knots.

As the power of the cruise-engine increased, however, there was increasing advantage in adding it to the full-power total, in a COGAG arrangement, rather than letting it stand idle. Moreover, if the disparity between the cruise and boost engine were sufficiently reduced, it would be possible to do away with the two-speed gear. The marine gas turbine, with its free power-turbine, is a flexible machine, which can still deliver nearly maximum power, at nearly maximum efficiency, over quite a wide range of speeds.

COGAG machinery, using four identical Olympus, was chosen for the new aircraft carriers at the end of 1969; the United States Navy adopted the same system a year later for their first gas turbine-powered destroyers, the *Spruance* class, using four General Electric LM-2500 engines. In the case of these two classes of ship, it would be possible to cruise on one engine with good economy, but to use similar engines in a COGAG system for a much smaller frigate or destroyer would be expensive in fuel. Although the power-requirement for such a ship at 24 knots is around 25,000hp, at 18-20 knots this is reduced to 10,000-15,000hp. In this power-range, the Olympus is uneconomical and even an engine such as the RB-211 is not ideal. If speed is reduced further, the comparison between the big engines and a good cruising engine such as the Tyne becomes more pronounced.

For a big ship such as the 20,000-ton *Invincible*, four Olympus engines provide a good machinery fit. For a 4000-ton destroyer, requiring half the power, four engines of half-Olympus size would provide the ideal answer, as long as they had a good enough efficiency to give really economical cruising power.

This was the half-size 'building block' for which a need had been mooted in 1968, to fill the gap between the Olympus and Tyne. It would enable almost any power-level to be provided for new ships. In smaller frigates or corvettes, hitherto powered by a single Olympus and one or more diesels, with a single propeller or a split drive, it would be possible to use two intermediate engines, either with cruising diesels or, if the efficiency were sufficiently good, by themselves in an extremely simple one-engine-per-shaft arrangement. With suitable gearing it would be quite feasible to use three engines driving twin propeller shafts. If higher power were required, it would be possible to mate two 15,000hp engines with two Olympus in a COGAG system and still have good cruising performance on the smaller engines alone, without having to resort to a two-speed gear box.

Within the range of established British aero-gas turbines there were two possible engines which might be considered as candidates for

One of a large aero-engine family. The Spey turbo-fan engine.

development into a 15,000hp marine unit. The first, and most obvious, was the Avon, which had an immense background of experience, was of about the right size, and had been in service as an industrial engine since 1964. It would have been an easy engine to convert to marine use, since it had already been adapted for operation at sea-level, in a polluted environment, and little further work would have been needed to produce a suitable gas-generator, although a new marine power-turbine would have been required.

The new marine gas turbine was to be used as a cruising engine, however, and a first requirement was that it should have the best possible efficiency, equal, at a lower power, to that which was attainable with the RB-211 at 27,500hp. The Avon was an old engine, and although it had been developed over the years, and would continue to be developed to higher powers in its industrial version, it had a relatively high fuel-consumption, and there was no further scope for any significant improvement in this respect.

The other possible candidate was the Spey. This was a more modern engine than the Avon, though still with an impressive aero-lineage and experience. It had a high compression-ratio, and could achieve the good efficiency that was a prerequisite. In its later versions, it was

also of the right size for the Navy's requirements.

In August 1972, the Ministry of Defence awarded a design-study contract to Rolls-Royce for a Marine Spey. It was to be some years before the final decision to carry out the full development was taken, and a good deal longer before the Navy would have the benefit of the new engine at sea. This was in part due to the Navy's perennial shortage of money to meet its commitments, and to the fact that the existing machinery still served all the existing ship-requirements very well. Despite some delays, however, engine development did proceed ahead of new ship design, because it was now appreciated that, however extensive the aero-background to a new marine engine, the process of re-design, development, and testing, up to the point where a unit was ready for service, would take longer than the detailed design and construction of a new class of ship.

Logically, therefore, having decided on the right specification for a new marine engine to fill the gap in the existing range, and thus meet the needs of a wide variety of ships, the right way to tackle the work was to go ahead with the development of the engine in advance of final ship-specification, so that it would be available, fully tested, when it was needed for the Fleet.

THE INTERMEDIATE ENGINE

The Speys were another of those prolific gas turbine families. The first Spey started test-bed running at the end of 1960, and the engine went into civil airline service, in the Hawker-Siddeley 'Trident' in 1964, with a thrust of 10,000lbs.

By 1972, when the Navy first came to consider it as a serious candidate for marine use, members of the now burgeoning family were in service, or about to go into service, in an impressive number of aircraft, including the 'Trident', BAC-111, Fokker 'Fellowship', and Grumman 'Gulfstream', in the civil sphere, as well as the Hawker-Siddeley 'Buccaneer' and the McDonnell-Douglas 'Phantom' fighters, and the 'Nimrod' maritime reconnaissance aircraft. Over 3000 Speys had been produced, and they had amassed, between them, some 8 million hours of flying. The latest and largest member of the family, known as the TF-41, had a thrust of 14,500lbs, and powered the new American A-7 'Corsair II' fighter.

This was an impressive background, but although it would obviously be of value to a naval development programme, it would not provide direct backing in some areas, because the aero-engine configuration made it necessary to carry out some major surgery, in addition to the more normal process of 'marinisation', to complete the conversion.

The Spey, like most of the newest engines, was a turbo-fan, and it used the well-tried two-spool type of compressor common to the Olympus, Tyne, and RB-211. Unlike the RB-211, however, the fan or 'bypass' element in the engine was represented, not by a separate fan on its own independent shaft, but by an extension to the blading in the first three stages of the low-pressure compressor, which thus protruded beyond the limits of the main compressor casing, and passed some of their air over the outside of the engine, through an external duct.

To convert this turbo-fan into a two-spool marine gas generator, supplying hot gas to a new design of power-turbine, meant getting rid of the fan element by 'cropping' the blades and re-designing the engine casings, so that the whole air-flow was directed through the interior of the engine to the combustion chambers. This was, in fact, a very simple and surprisingly effective operation. It could be done, literally, by cutting off the top half of existing rotor blades. New stator blades and casings would be needed, as well as re-matching of the turbines, but it could produce an extremely efficient engine for industrial applications, where the engine would be run, for most of the time, at its maximum design power.

For marine use, however, changes in material were required, to improve resistance to salt corrosion, and some further re-design of the low-pressure compressor blading was desirable in order to achieve maximum efficiency, both at maximum output, and at the intermediate powers which would still be required when cruising at reduced speed.

During the long discussions on future engine requirements, the power of the intermediate engine had increased. The Royal Navy had also 'gone metric', and what had started in the 1960s

*The largest of the Spey engine family powers the U.S. Navy's A.7 **Corsair** aircraft.*

as an engine of some 10,000hp was eventually defined as an 11MW (or 14,750hp) unit. At this power, the Spey would need to be able to operate for extended periods as a cruising engine, and be capable of long running between overhauls. It must have the best possible efficiency, both at full power and at part-load; it must have a smoke-free exhaust throughout the power range, and it must be designed to meet all the usual naval requirements in the matter of corrosion-resistance, shock -and blast-resistance, nuclear fallout-protection, quick starting, and rapid handling.

To produce the required power, and provide a margin above this, both to ensure long life and to allow for further development, the aero-progenitor selected was the TF-41, which had the advantage, not only of being the most powerful and most recent version, but of having a larger flow of air through the core of the engine (as distinct from the fan air flowing through the outer duct). This would enable the full cruising-power of 14,750hp to be achieved without re-design of the high-pressure spool, and without exceeding safe temperature limits in the high-pressure turbine. In fact, it would give a

maximum power of 17,000hp, with the capability of further development to a possible future naval requirement of 14MW, or 19,750hp, without any change in the basic layout of the engine.

Apart from the need to remove the by-pass element of the aero-engine, the process of 'marinisation' followed familiar lines, with strengthening of casings and bearings, modification of the combustion system, and changes in material. There was much experience to draw on. New material for turbine rotor-blades had been tested in the Marine Tyne; another new material for the turbine stators had had extensive running in industrial versions of the Olympus, operating in the highly salt atmosphere of the Arabian Gulf coast; a new design of combustion-chamber, which produced remarkably efficient and smoke-free burning of diesel fuel, was developed from research work started in the aero division of the company.

One innovation, which marked a departure from previous practice with the Olympus and Tyne, was the use of solid state electronics in the engine control system. Gas turbine controls have to take account of a number of different inputs — speeds, temperatures, and pressures —

and provide a number of safeguards to ensure completely reliable operation. Even in the far-off days of the early marine development, automatic controls had been a feature of the new engines that signalled a considerable advance over post-war steam machinery. When the Engineer-in-Chief of the Navy, visiting Bristol in 1956, first tried the controls of the Marine Proteus on the test bed, he was heard to exlaim, in admiration: 'I should think even a Marine could work this!' Gas turbine control systems have indeed been described, rather unkindly, as not only fool-proof but sailor-proof. The further advance to electronic controls was nevertheless viewed in some quarters with misgiving. Here again, however, experience with other engines and in other fields provided some reassurance, for apart from the Navy's growing use of electronic systems elsewhere, Rolls-Royce had by this time had considerable operating experience with electronic control systems on industrial gas turbines.

As with other marine gas turbines, the engine controls would be integrated with the general ship-system, so that the whole operation and monitoring of the machinery would be undertaken from the Ship Control Centre, and manoeuvring could be carried out direct from the bridge.

A new power turbine would be required, and as the Spey, like the RB-211, was a high compression-ratio engine, a two-stage turbine would be needed, to cope with the higher pressure drop between the gas generator outlet and the power-turbine exhaust.

When the general layout of the mounting system and the complete engine module came to be considered, a critical examination was made

RIGHT AND BELOW: *"... remarkably few changes .."*. *The marine Spey had to meet the same warship service requirements as the Olympus, and the designs of the two engine modules were very similar.*

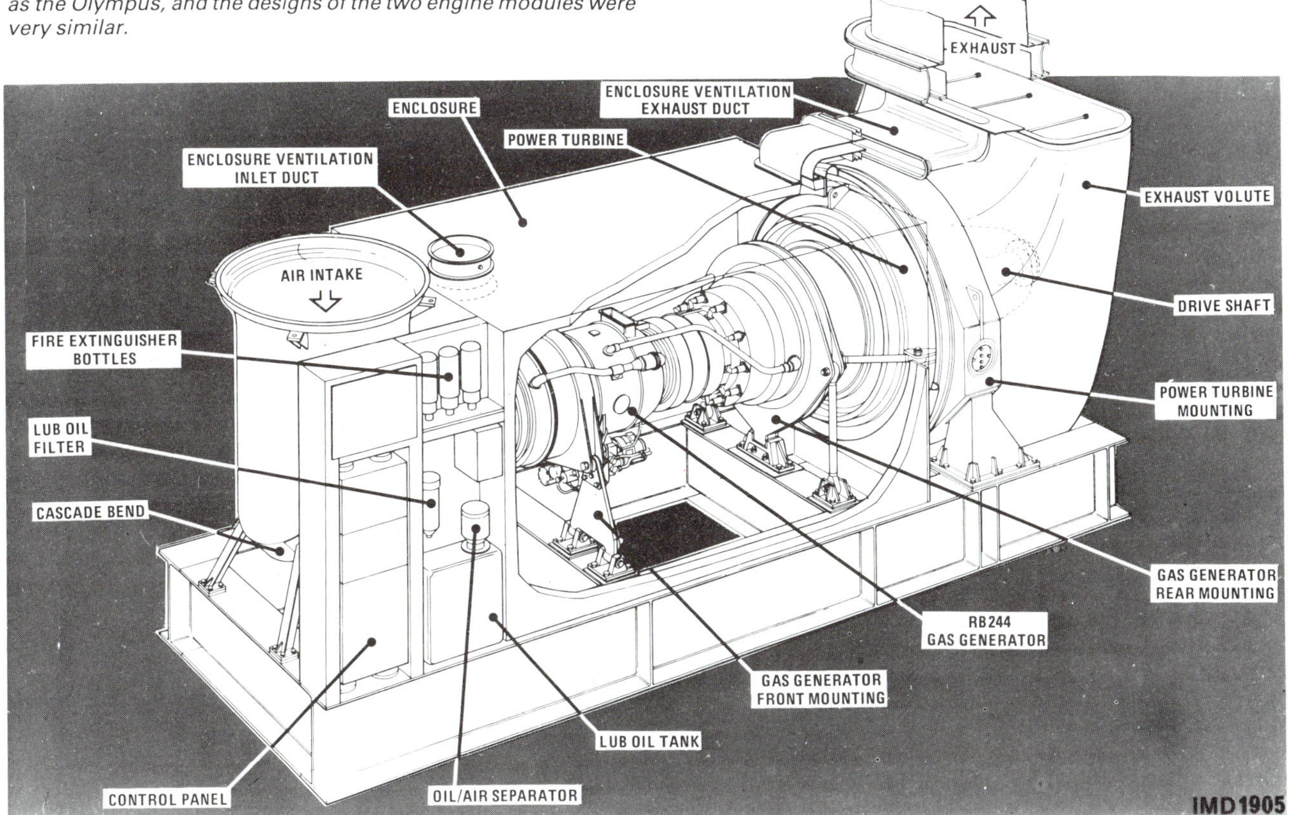

of the design philosophy on which the earlier modules had been planned. Remarkably few changes were made to the design which had been evolved eight years earlier for the Olympus. Alterations were made to the power turbine mountng system, to provide support for the shaft direct from an outer mounting ring, through the exhaust volute, rather than by a more circuitous route, via a pedestal at the rear of the volute. This gave a more rigid alignment between the power turbine rotor and its stator casing, without losing any of the advantages of plain bearings, clear of the hot gas stream, and accessible for inspection. The outer mounting-ring itself was carried off the base plate on trunnions, in the manner of a gun, as the Marine Proteus had been, 20 years earlier.

Changes were also made in the gas-generator mounting arrangement. This was now supported on a solid base-plate running the full length of the engine, which also carried the outer steel enclosure, but the effect of shock was reduced at the forward end of the gas generator by an ingenious front mounting which was offset to one side, so that the force from an underwater explosion would be felt as a twist in the engine carcase (which was very strong in torsion), rather than as a direct punch on the bottom of the compressor casing.

These were minor improvements in the mounting system: the main features of the unit — the overall shock-reinforcement to withstand 50G; the substantial long-life power-turbine; the ducted intake; the heat -and sound-insulated external casings, and the built-in auxiliary systems and controls — were virtually unchanged.

The initial design feasibility study was completed in 1973, and was followed by a detailed design contract, but while this was still in progress, industrial versions of the engine were already being built, and it went into service as a gas pumping engine in Canada in 1976, the first of the new high-efficiency engines designed as successors to the familiar Avon, which had started its industrial career just 12 years earlier.

We have already seen that industrial and marine requirements are by no means the same, and the gas pumping operation of the Spey was not going to remove the need for a marine development programme; but it did mean that much of the ironmongery was al-ready benefiting from sea-level running experience, and it did save money. When, finally, in the spring of 1977, the decision was made to go ahead with full development of the Marine Spey, three separate problems which would each have had a direct effect on the marine engine, had already been encountered in industrial service and solved.

The detailed design and development of the marine engine continued through the ensuing three years, with the usual rig-testing of engine-components, trials of three development gas-generators, and final running of the complete engine module; and it was not until the latter part of 1980 that this stage was reached.

At this point, the engine still had no home, as plans for new ships had not yet crystallized. There was, of course, still further work to be done on the engine, with a programme of test-bed endurance running, both in Rolls-Royce and and the National Gas Turbine Establishment, to build up experience and running hours, under simulated marine conditions, in advance of the engine going to sea.

Meanwhile, on the ship front, a number of new designs were being considered. In the RN, the first application planned was the engining of a Type 22 frigate with Marine Speys in place of the two Olympus engines. Such an arrangement, with two Tynes and two Speys, reduces maximum speed slightly, but this has relatively little effect on normal operational capability, and the ship can operate in the 18-25 knot speed-range with far greater fuel economy — indeed this machinery arrangement gives the best possible efficiency throughout the whole speed-range.

The much smaller Type 23 frigate, announced in mid-1981 as the Navy's next class of new surface ship, can be powered by one or two Speys in a CODOG or COGAG configuration. The twin-Spey arrangement lends itself well to the type of machinery layout which, as we saw earlier in considering the *Leander* class frigates, could provide a very simple and very efficient propulsion plant for a ship of this size, displacing around 2000 tons, with twin shafts, each driven by a single Spey through a reversing gearbox. It is possible to achieve some saving in capital cost by using two Speys with a combining gear-box in a single shaft arrangement, and this also produces some improvement in single-engine

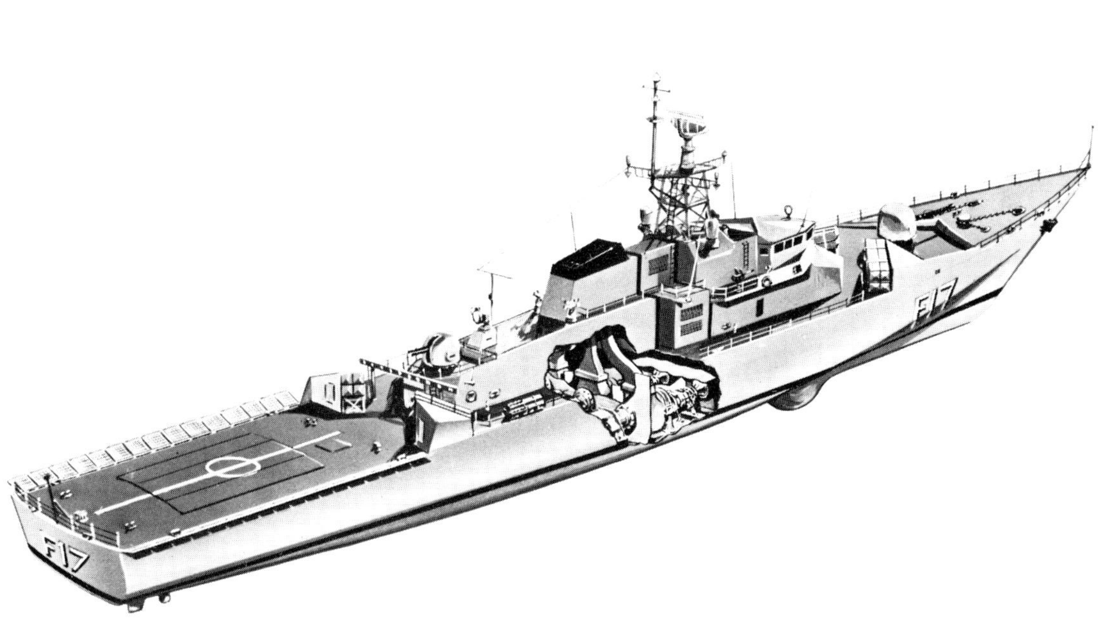

The Vosper Mark 17 frigate. Three Speys in a 1700-ton hull provide sparkling performance, with very good flexibility and fuel economy.

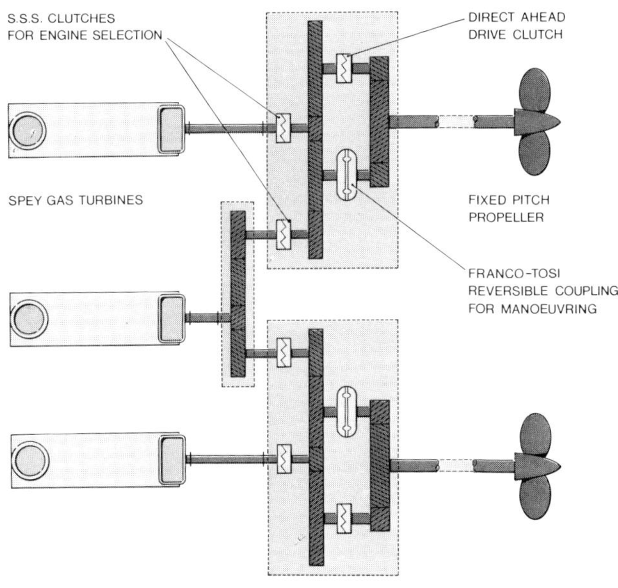

"Three engines driving two shafts allows operation on one, two, or three engines without having to resort to shaft-trailing. In this arrangement, fixed-pitch propellers are used, and reversing is by means of Franco-Tosi reversing gears."

cruising-efficiency, but at the cost of a reduction in manoeuvrability and in reliability (since damage to the single propeller or main reduction gearbox would immobilize the ship), and some increase in ship's draught through using one larger propeller in place of two smaller ones.

Perhaps the most interesting application, because it best demonstrates the advantages of the new intermediate engine in the typical modern frigate, is the three-Spey configuration proposed by Vosper in their Mark 17 frigate. This is planned as a twin-shaft arrangement, with each shaft driven through a separate reverse-reduction gear. The wing engines drive into the main reduction gearboxes on their respective sides, and the centre engines drive into both gearboxes through its own central splitter gear. For full power all three engines are connected; for cruising up to 20-25 knots, the centre engine drives both shafts; for intermediate speeds, and for manoeuvring, any two of the three engines can drive the two propeller shafts independently, either in direct ahead-drive, or through the reversing gear.

This has the advantage not only of providing the best overall fuel-economy — though a single Tyne would be even more economical at cruising speeds below 14-16 knots — but also of affording the best reliability in the case of damage to any part of the propulsion system.

In the case of the Mark 17 design, the three-Spey machinery has been applied to a high-performance ship of some 1700 tons, with a maximum speed which must be approaching 35 knots. Despite her small size, her designers have managed to mount an impressive armament. This has been done partly by employing a new lightweight 'Sea Dart' missile system, and partly by reducing complement through the use of extensive automation. As a result, in addition to the 'Sea Dart' SAM (which also has a surface capability), the ship can carry the OTO-Melara 76mm gun, twin Breda 40mm., and a 'Lynx' helicopter equipped to carry ASW torpedoes.

Gas turbine propulsion-plant, integrated into a digital ship-control system, makes some contribution to this good result. We have already seen how the transition from steam machinery to gas turbines cut down on machinery space requirements and, to a greater degree, reduced engine-room complement — and thus saved not only on the cost of manning, but on the cost of, and space occupied by, crew accommodation. The transition from the Olympus-Tyne machinery of today's frigates and destroyers to a three-Spey installation makes little difference to manning requirements, but it certainly reduces the space occupied by machinery. The three Speys will produce, today, slightly less power than the twin Olympus-Tyne arrangement — a nominal 51,000hp against 56,000hp, or a reduction of just under 10 per cent — but the saving in engine space, and in the area occupied by uptakes and downtakes, is 30 per cent.

When we consider the comparison between steam and a two-Spey gas turbine installation in frigates such as the *Leander* class, with an installed power of 30,000hp, the contrast is even more striking. The space occupied by the main propulsion machinery is virtually halved, and at the same time the fuel-stowage for the same endurance, is reduced by some 30 per cent. In ships of this type, the gas turbine does make major improvements.

At the other end of the scale, in the new guided missile destroyers now being built for the Japanese Maritime Self-Defence Force, with a displacement of over 5000 tons, neither the standard Olympus-Tyne COGOG installation, currently being used for the Japanese destroyers, nor a four-Spey COGAG arrangement would fully meet the requirements. The cruising speed on two Tynes would have been limited to 15 knots or less; the maximum speed, even with four Speys, would, in this particular case, be rather under 30 knots. However, the combination of Olympus and Spey, which was eventually chosen, will provide a maximum speed on all four engines of more than 30 knots, and a cruising speed on two Speys approaching 24 knots, with the best available fuel-economy.

So the three engine 'building-blocks', between them, can provide machinery for as wide a range of corvettes or frigates or destroyers as we can envisage at present, from under 1000 tons up to 6000 tons or more.

THE FUTURE?

oretelling the future is a highly dangerous business. In 1945 Admiral Leahy told President Truman: 'The atom bomb will never go off, and I speak as an expert in explosives'. In the last century, the great physicist Lord Kelvin declared that 'radio has no future', and a little earlier that eminent scientist, Professor Lardner, exclaimed: 'Men might as well expect to walk on the moon as cross the Atlantic in one of those steamships!'.

The steamships are disappearing; the gas turbine ships may follow them, for there have been plenty of prophets during the last 50 years who have been ready to forecast the end of all ships. However, if we assume, merely for purposes of argument, that the world continues to need warships to maintain the peace at sea for some years to come, there are two factors affecting propulsion machinery which are sufficiently obvious to deserve comment. In the first place, warships are undeniably getting more expensive, in real money terms; second, the cost of fuel is also increasing, and although the scaremongers of a few years back appear to have been confounded, it is undeniable that the oil will run out some day.

The increase in ship-costs is staggering. It is, perhaps, unfair to go back to the days of the sail, but if we take the 19th century steam gunboat as the forerunner of today's frigate (for the lineal descendant of Nelson's frigate was the cruiser), we find that in 1855, a 1000-tonner cost £17,000, but by the 1890s this had increased to £64,000. In the 1930s, with the 'A-to-I' destroyers, vastly more powerful ships of 1500 tons, the cost had gone up to £375,000 — or say £150,000 in 19th Century money. Forty years later, in 1970, a *Leander* frigate displaces nearly 3000 tons and costs £17 million — but we probably have to divide by ten to convert to 19th Century values. Today, a Type 22 of 4000 tons will cost £90 million — or some £3 million in 19th Century money; so our cost-per-ton has gone up from £17 to £750 in real terms.

Of course, you get vastly more hitting power and sophistication for your money, and the largest single item in today's bill — electronics — did not figure in the 19th Century accounts, but the bill must still be paid. This helps to explain the fact that, while individual ship-types — cruisers, destroyers, and so forth — continually increase in size over the years, the largest types become, each in turn, too big and unwieldy to sustain, and disappear like the dinosaurs, to be replaced by much smaller types growing from below. Thus the battleships and battlecruisers are being followed by the carriers and cruisers into the pages of the history books, but the wartime MTB has grown into a 400-ton fast attack craft with the same sort of surface and AA armament, on a smaller scale, as the contemporary frigate — the German Type 143 carries 'Exocet' SSMs, the 76mm OTO-Melara gun and two 21 inch torpedoes, with a new SAM under development.

Such craft as these cannot compete, in endurance, seakeeping or autonomy, with large ships in open ocean conditions, and it is difficult to see them in an antisubmarine role; but they are well able to cover a wide range of operations in European waters, or in other more sheltered areas around the world. When we turn to unconventional craft such as hydrofoils, the Italian *Sparviero* class, displacing only 65 tons, are surprisingly seaworthy, and still manage to mount a 76mm gun and 'Otomat' SSM, while the Royal Navy is now experimenting with hydrofoils as a possible type of future offshore patrol craft capable of a North Sea role.

We can see the same tendency displaced within a single group of larger ships, in the RN frigates and destroyers of the past ten years, with the Type 82, as it were, outgrowing its strength, to be replaced by the Type 42 and Type 22, which, in their turn, now give way to the Type 23, less than half their size.

These step-changes in ship-size have been made possible without too great a falling-off in fighting capacity by the introduction of successive lightweight missile systems, by the use of micro-electronics, and by a great increase in automation and repair-by-replacement, which in turn has led to much reduced manning. There is no doubt that the gas turbine has contributed significantly to the process, both by economising in weight and space, and perhaps even more by lending itself readily to automation, and needing virtually no onboard maintenance.

If this interpretation of present trends is accepted, then, in the first place — leaving aside the fuel question, which we will come to later — there is a continuing case to be made for the gas turbine, *vis-á-vis* the lightweight diesel engine. This should apply in any ship in which the unit power-size for the propulsion machinery is above 5000hp. Below this size, the gas turbine starts to become very much less competitive, both in capital cost and in running costs. Patrol boats gain less from reduction in maintenance since they cannot, in any case, operate without base support, and a fast attack craft, with no ASW role, is not going to be worried about underwater noise.

In the second place, one way in which equipment weights, and hence ship size, can be reduced considerably is by relaxing the requirements for shock-resistance. Such requirements do not, in any case, normally apply to fast patrol boats and other small craft, and as far as gas turbines are concerned, they make a considerable difference. The standard Tyne cruising engine module weighs 31,000lbs (14,061Kg.) compared with some 2700lbs (1225Kg.) for the lightweight version of the engine, though this is not quite a fair comparison, as the cruising Tyne includes a heavy reduction-gear.

In the case of the Spey, where we are comparing like with like, except for the outer enclosure and the intake cascade bend, the difference is between 42,500lbs (19,300Kg.) for the big ship module and 18,000lbs (8200Kg.) for the standard lightweight version of the engine. It should be noted that this reduction can be achieved without any internal alteration to the engine, so that its performance is unchanged.

For ocean-going ships, such weight-saving is of less value, accepting that the North Atlantic is a pretty uncomfortable place in anything under

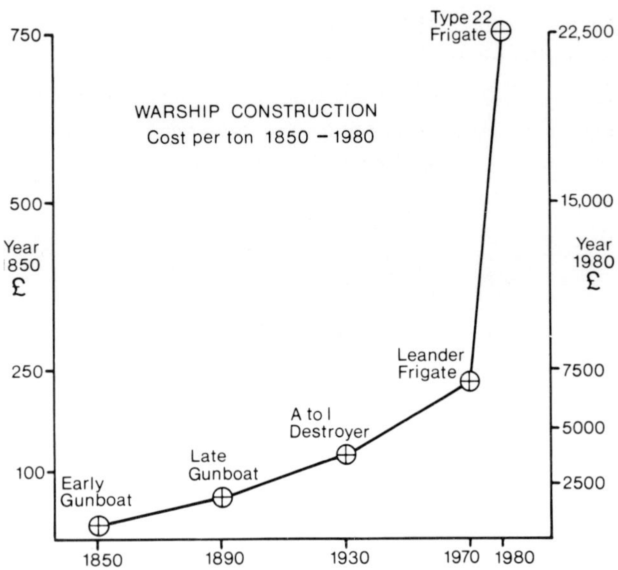

The increase in ship costs since the mid-nineteenth century. Electronics have made the biggest contribution, but perhaps the most daunting figure is the trebling of ship costs in the last ten years, which, with the collapse of the pound, has produced a tenfold increase in the price per ton.

about 1200 tons — e.g. 'V & W' destroyers, the American four-stackers, and 'Flower' class corvettes — and there is in any event no way of providing sufficient endurance in a smaller hull.

This brings us to the other factor which affects the choice of machinery: fuel cost and availability.

There have been so many and varied forecasts of long-term fuel supplies that it is fairly easy to pick one which supports almost any argument. If oil, gas, and coal run out, then — if we rule out solar energy and oars — we are left with nuclear power or sail. Nuclear power could well use gas turbines for propulsion, in place of the present steam turbines, and studies have shown that such a system could effect considerable economies in weight, space, and cost. However, all the calculations that have so far been made seem to conclude that, whatever the price of oil, nuclear propulsion is unlikely to compete except in larger ships, above 7000-8000 tons, and only then if a reasonable number are built, to offset initial high development-costs; so until the oil and coal do run out, nuclear propulsion seems unlikely to play a major part in most navies, other than in submarines.

In parenthesis, it may be noted that, although there is a school of thought which would maintain that the submarine can do anything the surface ship can do, and better, it is still reasonable to assume that the world's navies will continue to build some surface ships for the foreseeable future. The submarine is singularly useless in combating attack from the air or from light surface craft, and not conspicuously suitable for supporting amphibious operations, while it is capable of performing hardly any of the navies' normal peacetime or peace-keeping tasks.

To return to our fuel: clearly the oil will run out in time. In 1973 it was calculated that, at the growth-rate in consumption of 7½ per cent per annum which then obtained, world oil, including estimated new discoveries and difficult sources such as tar sands, would run out in 2008. Obviously, however, this could not happen, since oil-production could not possibly keep pace with such a growth in demand, and would, indeed, inevitably decline, as supplies became harder to win. Since 1973, consumption has dropped and the whole position has changed; 7½ per cent growth-rate now seems highly unlikely, and the prophets of doom are keeping their heads down. The oil will last a little longer, and coal, and oil derived from coal, will carry us forward for some time beyond the end of the Oil Age.

It is necessary to bear in mind, nevertheless, that the time-scale which we have to apply to ship-machinery is itself a long one. The new ships being conceived today will be built during the 1990s, and some may still be in service after 2030. Similarly, if new engines were being planned today, they would probably still be running well after 2030. To change the present fuel — for example to nuclear fuel — would equally take a long time — and cost a great deal of money. It would be possible, of course, when the oil runs out, to revert to coal, with much less capital expenditure than would be required to develop nuclear propulsion, and it would even be possible to use coal-fired gas turbines — experimental running has already been done, by Rolls-Royce, amongst others. But the trouble about switching to coal is that it is a far more bulky fuel that oil, and more difficult to stow and handle, so that ship-size would have to be increased by some 50 per cent to accommodate the change, increased manning would also be needed, and the costs would go up accordingly. It would almost certainly be more cost-effective to continue with oil-fired machinery and use coal-derived oil, which is already on the way to becom-

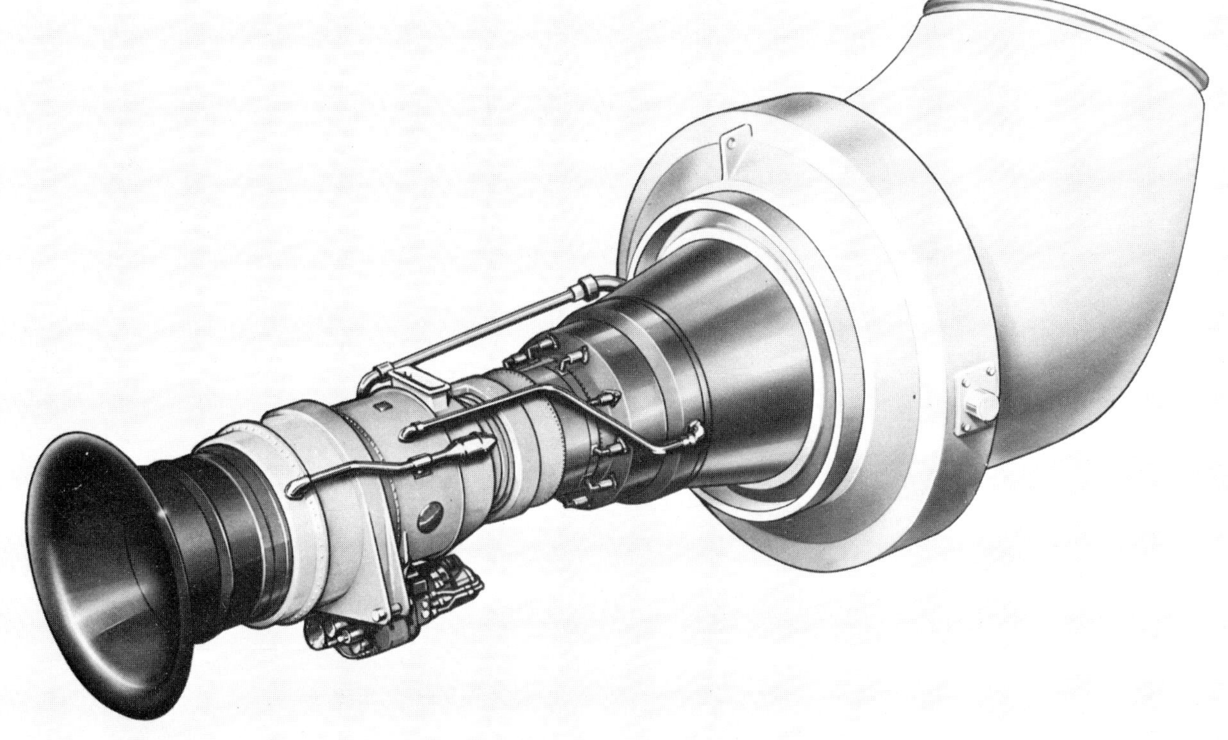

The lightweight version of the Marine Spey. Although the basic engine is unchanged, the mounting system and installation can be drastically simplified, and such an engine could still have a role in the small frigates and corvettes of the future, as well as in larger high-speed craft.

ing competitive in price with petroleum-based distillate.

In spite of the warnings which we sounded at the beginning of this chapter, therefore, it seems a safe bet to say that oil, in some form, will still be available beyond the life of the next generation of ships, and beyond the life of anyone reading these words today.

It needs to be remembered, as supplies start to decline, that naval fuel consumption represents only a tiny proportion of a country's total energy consumption. It is difficult to generalise about this subject, as figures are not readily available, but we can make a rough estimate of fuel consumption for the RN surface fleet of frigates and destroyers — i.e., the classes using gas turbines, which make up the greater part of any country's fleet. Some 50 such ships are currently in commission, leaving out those refitting or building, and assuming 2000-2500 hours steaming a year, we might expect annual fuel-consumption to average not more than 5000 tons per ship. This gives a total annual fuel-con-

sumption of 250,000 tons. But the British national energy-consumption, in tons of oil equivalent, is over 200 million tons a year, so the consumption of the gas turbine (or potentially gas turbine) fleet represents only one-eighth of one per cent of the national total.

If we can assume that the navy will continue to burn oil, this does not necessarily mean that it will continue to use gas turbines, or today's sort of gas turbines, as its main means of propulsion. A return to steam is unlikely, because it would not only re-introduce the complexity and high burden of maintenance which we have been at pains to get rid of, but it would cost more. Present-day gas turbines, such as the Spey, can achieve an efficiency throughout the ship's operating range which is better than that for current steam plant. Admittedly we need refined fuel for the gas turbine, which is more expensive than traditional furnance fuel; but most of the world's navies were already using diesel fuel for boiler-firing in order to reduce boiler-maintenance, even before gas turbines

"Combined cycle. The ultimate in fuel economy, and much better than the diesel; but it could bring back a lot of the old maintenance problems."

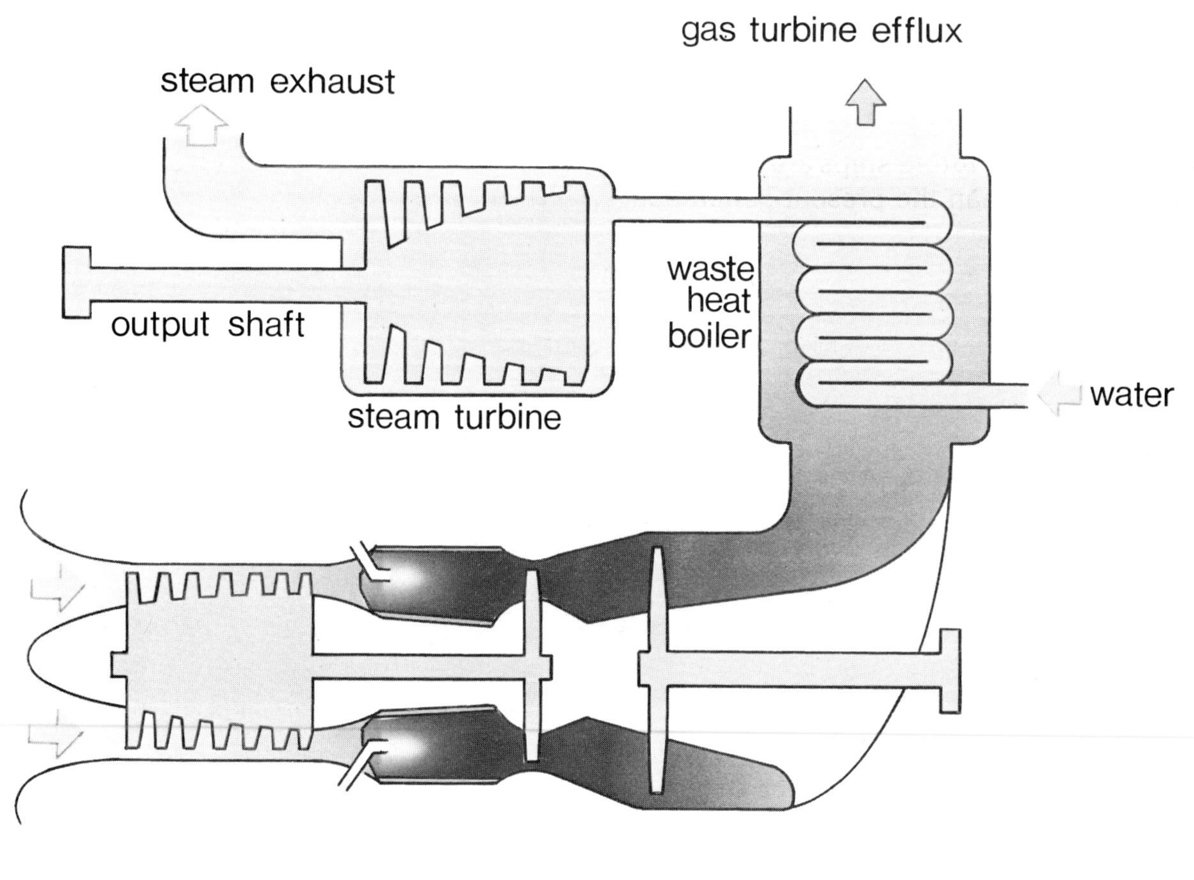

were introduced — the Royal Navy made the change during the 1960s — and there seems little likelihood of their reverting to residual fuel.

Using diesel engines, either as the sole means of propulsion or in CODOG installations, does secure a saving in fuel, and even if this will have little effect from the point of view of our dwindling fuel stocks, it still needs to be taken account of from an economic point of view. If we take the example of a three-Spey frigate, such as the Vosper Mark 17, the saving achieved by using all-diesel propulsion would be, on an average, around 15-20 per cent, but to secure this would mean changes in ship design, to install the diesels and to accommodate additional engine-room staff, so that the resulting increase in ship size could to some extent cancel the diesel advantage; but this is not a simple matter, and has to be worked out in each individual case. On the basis of today's economic forecasts, it seems to have been agreed by most countries that the balance of advantage, even where underwater noise is not a prime consideration, lies with the gas turbine for ships over about 500 tons and for specialized high-speed craft.

Reducing the gas turbine's fuel disadvantage to manageable proportions does necessitate choosing the right engine arrangement for the size of ship and the expected operating profile. In this respect it has become clear that, always accepting that future ships are likely to be at least no larger than the present generation, the decision of the Navy to choose the 15,000hp Spey rather than the 30,000hp RB-211 — bearing in mind that an Olympus was already available for big ship boost-purposes — was a good one, and a unit of this size is probably about the nearest we can get to a universal warship prime-mover.

It is possible to reach higher efficiencies with gas turbines than are attainable either with conventional steam or with diesel machinery, by using a combined-cycle arrangement, utilising the heat in the engine's exhaust gases to raise steam for a steam turbine, which adds to the total power available without increasing fuel consumption. In this way, output can be increased by some 30 per cent, and efficiency can go up from the present 35 per cent of a Spey or RB-211 to as high as 45 per cent. But this again means bringing back steam. It is, perhaps, significant that the United States Navy, the navy with much the greatest technical resources as well as the greatest operating-range requirement, is the only one which seems at present to be seriously studying combined-cycle plant for warship propulsion.

With or without combined-cycle plant, the propulsion system would still be based on the gas turbine, or on a gas turbine-diesel combination, and there seems to be no reasonable alternative in sight at present, until we are all driven back to the appallingly labour-intensive, but much more civilised and quite inexhaustible resource of sail.

GLOSSARY

ENGINE TERMS

Compressors   All the compressors in the propulsion gas turbines here discussed (with the exception of the H.P. compressor in the RM 60) are of axial type. Many earlier gas turbines used centrifugal compressors — the Rolls-Royce Dart is a good example — and they are still common in small engines. They have the advantage of being very rugged, resistant to foreign object damage (q.v.), and less sensitive to the effects of compressor stall (q.v.). Their disadvantage, in aero engines, is that they have a large frontal area, causing high drag, and a multi-stage centrifugal compressor, which requires large casings, and in which the air has to be turned through 180° between each stage, is both bulky and relatively inefficient.

In the axial-flow compressor, the air-flow is much more direct, there is less frictional loss, and the frontal area is far smaller. For these reasons, virtually all the larger aero gas turbines, from which current marine engines are derived, use axial compressors; but these do suffer from the disadvantage that the unsupported compressor blades are rather more vulnerable to foreign object damage or vibration.

Compressor stall   The breakdown of the air flow through a compressor can be caused in a number of ways. Internally, it can result from mismatching of the front and rear ends of the compressor when it is not running at its design point. This can be avoided by using a two-spool compressor (where the relative speeds of the two spools vary automatically to meet the demand), or by fitting blow-off valves (which release some of the air from the front stages of the compressor when it is running at low speed), or by variable stator blades (which can be rotated, to match the reduced air-flow through the front stages, at low speed).

Stalling can be caused, externally, by obstruction of the air intake, or by excessively violent acceleration of the engine. It is also more liable to occur if there is a heavy build-up of salt on the compressor blades.

It may take the form of a rapid succession of stalls — for this reason, it is also known as

compressor 'surging' — and the violent fluctuation in the air flow and pressure can result in blade failure. It can also, through the cutting off of air to the combustion chambers, cause a sudden disastrous rise in temperature, with consequent burning of turbine blades, if the fuel system is not designed — as modern systems are — to cope with this contingency. Both mishaps occurred in earlier marine engines.

Development   There are numerous references to development, and to its high cost. As understood in the gas turbine world, this is a term which sometimes puzzles laymen, and even honest marine engineers. The need for development stems from the fact that the gas turbine designer is working at the limits of his knowledge. If all the answers were known, an engine could be designed, and built, and installed, and that would be the end of the story — and of development engineers. Fortunately for the latter, we are dealing here with advanced and complex engineering where no men, or computers, can foresee all the results of a design decision. For obvious reasons, engines are designed to produce the highest possible power and efficiency, and temperatures, pressures and stresses are pushed to the limit, but we still need high reliability and long life. What will happen when an engine is started on a test bed is largely, though not entirely, predictable. What will happen after 1000 hours running, or in a slam acceleration, or in extreme conditions of temperature or environment must also be predicted, but is much less certain. Performance can only be verified, and action taken to improve it, by building engines, running them, possibly breaking them, modifying them, and running them again until they meet the full design requirements and all the answers, as far as possible, are known.

Of course it is possible to build an engine which will run satisfactorily without the need for any development, but this means designing to very conservative standards. The result would be a unit much more bulky and much less efficient, as well as less reliable.

Efficiency   As used in this book, efficiency is the ratio between the energy fed into the engine in the form of fuel, and the energy produced by the engine in the form of shaft power — in modern engines, 30-35%. The remaining 65-70% of fuel energy is dissipated in the form of heat, mainly in the exhaust, and to a very small extent in the lubricating oil or to the surrounding air.

In the marine world, it is common to express efficiency in terms of specific fuel consumption (s.f.c) measured in pounds of fuel consumed per horsepower-hour (or grams per metric horsepower-hour; or kg/kw-hr). Using the more arcane units of the Systeme Internationale, we ought nowadays to express s.f.c. in kilojoules per kilowatt-hour, but since this means nothing to most people, I have stuck to lb/hp-hr, or percentage efficiency, in this book.

Lb/hp-hr can be converted to grams/CV-hr, multiplying by 447, or to kg/kw-hr, multiplying by 0.608. With standard diesel fuel:

$$\text{Efficiency} = \frac{13.72}{\text{lb/hp-hr}} = \frac{6140}{\text{grs/CV-hr}} = \frac{8.35}{\text{kg/kw-hr}}$$

Foreign object damage   Damage, usually to compressor blades, caused by a gas turbine ingesting any solid matter through its air intake. In an aero engine, the damage is generally due to stones on the runway, or birds, or possibly ice. In the marine world, where the engine is buried deep in the ship and protected from the outside air, damage is rare, and will more often come from a loose bolt or rivet head, or possibly from tools inadvertently left in the downtake after refit.

Fuel   Aero engines run on kerosene, or what the uninitiated (in England) refer to as paraffin (AVTUR, AVTAG, and AVCAT are all types of aviation kerosene). Marine engines run on distillate diesel fuel — DDF, gas oil, Dieso, etc.

This is largely a matter of availability. Kerosene will meet the very varied conditions of aircraft operation, and since it is universally specified for aero engines, it is universally available at airports.

Distillate diesel fuel produces problems at high altitude, which need not concern us, and is also slightly more difficult to manage from the point of view of smoke, carbon deposition, and sulphur corrosion; but it is still a good fuel, and it is universally available from marine sources, because it is used in high-speed diesel engines and also, nowadays, for boiler firing in most navies.

The fuel used for boiler firing in commercial applications, and in the large slow-speed diesels — furnace fuel, or F.F.O., Bunker C, residual fuel — can vary enormously according to the origin of the crude oil from which it is derived. Some crude is very clean and light — North Sea crude can **almost** be fed into a gas turbine untreated, and Libyan is also fine stuff. Some is waxy — residue from the Australian Bass Strait crude is like brown boot polish but very pure; Nigerian is similar. Nearly all the rest contain lethal quantities of vanadium, which has a voracious appetite for turbine blades,

and unpleasantly high levels of sulphur, sodium, and sediment.

As a warship must have a reliable fuel, and must, at the same time, be able to embark fuel in any part of the world where its duties may call it, navies use distillate diesel, for which the specification is virtually the same world-wide, though heavy residual fuel has been used in some industrial and merchant navy applications in cases where the supply can be confined to a known local source.

GAS TURBINE CYCLES

Simple cycle   With the exception of the EL 60A and the RM 60, all the propulsion gas turbines referred to in this book — and all in service today — are **simple-cycle** engines, working on the basic principle described in Chapter I and illustrated on Page 6. The front section of this, comprising the compressor, combustion chambers, and compressor turbine, is the **gas generator** (because it produces the hot exhaust gas which provides the driving force) and, by itself, constitutes the basic aircraft jet engine.

Aero engines in which the hot gas drives an output shaft, rather than issuing as a jet, are described as **turbo-prop** engines (for conventional propeller-driven aircraft) or **turbo-shaft** engines (for helicopters).

By-pass or **turbo-fan** engine.   This is the common type of aero engine in 'jet' aircraft today. In this type, some of the energy in the exhaust is used to drive an additional turbine, behind the compressor turbine. This, in turn, drives a fan, which passes extra air round the exterior of the engine, to mix it with the exhaust gases issuing at the rear. By this means, a much bigger weight of air is shifted than is the case with the pure jet, but the air-plus-exhaust is travelling at a lower speed than in the pure jet. This gives improved efficiency at the subsonic speeds which are used by most passenger aircraft, as well as reducing noise.

The marine engineer wants **all** the available power from the 'jet engine', or gas generator, to drive an output shaft, and so he substitutes a **power turbine** for the aero engine's fan turbine and fan (or by-pass section of the compressor), and he removes the external duct, retaining only the inner core of the turbo-fan, which is, in effect, the gas generator section of a simple-cycle gas turbine.

Regenerative engine   In this type of engine, of which the EL 60A and the RM 60 are examples, the exhaust gas is passed through a **heat exchanger,** or **regenerator** (or **recuperator**), where part of its heat is transferred to the compressor discharge air, before it enters the combustion chambers. Some of the energy in the exhaust gas, which would otherwise be wasted, is thus usefully *employed to improve the efficiency of the engine.*

The benefit is more marked when there is a large temperature difference between the exhaust gases and the air from the compressor, and since the compressor air is, in any case, heated by being compressed, the regenerative engine gains most when the compression ratio is low. Modern simple-cycle engines, with a high compression ratio, have efficiencies almost as good as could be obtained with a regenerative cycle, at least when they are running at or near their maximum power.

The heat exchanger is a heavy and bulky piece of equipment, inclined to suffer from thermal stress problems, particularly when it has to cope with rapid changes of temperature due to manoeuvring, and the regenerative engine has not found much favour in naval circles, though it has been used in one or two merchant ship applications with varying success.

For obvious reasons, regenerative engines are not normally used in aircraft.

Temperature   *All talk of gas turbine performance is tied up with temperature. The hotter the gas, the more energy it contains, and can thus impart to the turbines — and the harder the problem of protecting blades from melting or corroding.*

Temperatures in the combustion chambers, where the fuel is burning, are far higher than the metal can tolerate, and the chamber walls are protected by ducting surplus air, not used to burn the fuel, round the outside of the chamber or 'flame tube'. More air is mixed with the hot gases before they are allowed to enter the turbines — hence the gas turbine's need for large intake and exhaust ducts, for the air used to cool the combustors and dilute the hot gases is about three times as much as that actually needed to burn the fuel.

Even with the addition of so much cooling air, the temperature of the gas entering the turbine is still, nowadays, too high for the metal to withstand, so the blades are cooled by having further air pumped through them, or bled over their surface in a protective 'skin'.

This is fine for an aero engine, but it has limitations when applied to marine gas turbines. The main concern in the aero engine is to limit overall metal temperature, since, if it rises too high, the blades lose their strength and start to elongate, or 'creep', under the stress of centrifugal force.

The mariner, however, is more particularly concerned with metal corrosion, caused by the combination of salt (in the intake air or the fuel) and the sulphur which is present in diesel oil. Combatting of corrosion depends, not on limiting overall temperature, but on avoiding dangerous surface temperatures. Internal cooling lowers the average temperature but has less effect on the external surface. External cooling involves bleeding air through minute holes in the surface of the

blade, which may become blocked by dust or salt under warship conditions. Even the best cooling will produce considerable variations in temperature over the surface of the blade.

Corrosion can vary enormously according to the amount of salt in the air, the amount of sulphur or salt in the fuel, the metal temperature, and the type of metal used. Materials which can best withstand high temperatures may be vulnerable to corrosion, and ensuring adequate corrosion resistance may involve some sacrifice in creep strength. Much can be done by treating blades with a protective coating. In earlier days, ceramic coatings were used, and were excellent for stopping corrosion, but too short-lived, and vulnerable to mechanical failure. A variety of metal coatings has been used, and some form of metal treatment is now standard practice.

Uprating   *Much has been said about uprating of gas turbines. This is one of the continuing results of development (q.v.) — the other being improvement in reliability and life — and it takes two forms. Firstly, an engine can be uprated — i.e., made to produce more power — simply by opening the throttle a bit further. If it was already stretched to the limit, this may break it — or at least seriously reduce its life; but engines are commonly introduced into service at power levels which err on the safe side, and experience may show, after a time, that some uprating is feasible without any structural re-design, possibly through the use of improved materials.*

The second form of uprating comes as a result of design changes — perhaps an extra stage on the compressor, perhaps the introduction or extension of blade cooling, or perhaps more radical re-design. Sometimes quite minor changes can give substantial gains. Before the marine Proteus went to sea, the power was increased by 8%, without any rise in temperature, simply by electro-polishing of the compressor blades and a small adjustment in the angle of the entry guide vanes.

SHIP TERMS

Availability   *As used of a ship, availability is the percentage of time for which the ship is available for service. Thus availability is reduced by the amount of time the ship is out of action either for routine maintenance and refit, or through the breakdown of equipment, or as a result of action or weather damage. So reduction in maintenance time, improvement in reliability, resistance to damage, and good seakeeping all improve availability.*

It will be appreciated, in the first place, that what with periodic refits, and a minimum of leave which even sailors need, and the inevitable break-downs which are bound to occur in such a complicated package of equipment, all ships must spend an appreciable amount of time out of action. Secondly, increased availability means that fewer ships are needed for a given task.

It is worth reflecting that, to maintain a single ship continuously on the abortive Beira patrol, a total of six frigates were required.

Coastal Forces   A rather vague term, which can certainly include inshore minesweepers and patrol boats, but which is commonly used as a generic word for small high-speed craft. The progenitor of the type was the Coastal Motor Board (CMB) of the first world war. From this came the Motor Torpedo Boat (MTB) and, purely as a counter to the MTB, the MGB, or Motor Gunboat. When, inevitably, the gun and torpedo roles were merged, the craft was known as a Fast Patrol Boat (FPB) to distinguish it from the ordinary (slow) Patrol Boat, which was altogether inferior, or the later Offshore Patrol Vessel, which was something quite different and very much larger.

In recent years, Coastal Forces have tended to be referred to as Light Forces and Fast Patrol Boats have become Fast Attack Craft (which sounds more aggressive and better value for money), or even Fast Missile Craft.

Displacement   Ship tonnage is confusing, because merchant ships use Gross and Net tonnage (which have nothing to do with weight, but are purely measures of cubic capacity, of the ship itself and the cargo space, respectively), while tankers are measured in deadweight tonnage (which is the weight of the cargo). For warships, we use displacement, which is meant to signify the actual weight of the ship; but we still confuse the issue by referring to 'standard displacement', normally signifying the original design displacement, without fuel and stores, and 'deep displacement', which is the final full-load figure — or the final officially declared figure — including fuel and stores. Since ships tend to grow during the gestation period, the difference between 'standard' and 'deep' may be a large one, and actual displacement can be considerably above the declared figure.

For the sake of uniformity, I have used published deep displacement figures throughout this book.

I have also followed Jane's Fighting Ships in using the old-fashioned 'ton' rather than the metric 'tonne'. I suspect that, in this case, the difference is purely in the spelling.

Horsepower   British horsepower has been used throughout this book, except where specifically stated. Where the horsepower of an engine is given, without qualification, the figure is the power at which the engine is run during its qualifying test, corrected to standard conditions of temperature and pressure.

In a ship installation, the power of the engine will be reduced, by a varying amount, due to the pressure loss in the ship's air intake system and the back-pressure in the exhaust system. There will be some further loss of power in the main reduction gearing and the propeller shaft, so that shaft horsepower (shp), or the actual power transmitted to the ship's propellers, is normally some 7-8% lower than the test bed power of the engine.

Major warships   Another rather vague term, traditionally comprising all ocean-going types of warship, from corvettes to battleships, and embracing, roughly speaking, anything over 500 tons. Purists might exclude ocean minesweepers and offshore patrol vessels (between 500 and 1500 tons), but as they are currently diesel-powered they are not of direct interest to us.

Operating profile   The proportion of steaming time spent at different speeds. This is a necessary piece of data for calculating endurance, fuel stowage, machinery requirements, and operating costs. It depends basically on staff requirements, though it will vary in practice with actual operational needs, and may be influenced by machinery characteristics and fuel cost limitations.

It can be expressed either as percentage time in discrete speed bands, or in the form of an 'S' curve, which plots, against ship speed, the percentage of total steaming time which is spent at or below each speed level.

Some time is obviously given to manoeuvring, or steaming at low speed in confined waters; some is spent at high speed during operations or emergencies; but 60% or more is normally spent at cruising speed, around 15-20 knots. In practice, full speed is very seldom used. A survey of the Royal Navy's operations during the last war showed that, even then, full power had been used for less than one percent of the total steaming time.

Steaming   This is something of a misnomer, but it seems to have become accepted, for running on gas turbines, rather as 'sailing' is accepted for leaving harbour on steam turbines. Some brave spirits tried to introduce 'gassing'.

ACRONYMS AND PROPER NAMES

Machinery arrangements   The introduction of gas turbines has spawned a whole series of acronyms to describe the various possible machinery arrangements:

CODOG   COmbined Diesel Or Gas turbine (one or the other)

CODAG   COmbined Diesel And Gas turbine (both can run together)

COGOG COmbined Gas turbine Or Gas turbine (separate engines for cruise and boost)

COGAG COmbined Gas turbine And Gas turbine (all engines can run together)

COSAG COmbined Steam And Gas turbine (the early compromise for big ships)

CONAG COmbined Nuclear And Gas turbine (not unattractive, though never used)

COGAGE COmbined Gas turbine And Gas turbo-Electric etc.

Other acronyms

ASW Anti-submarine warfare

SAM Surface-to-air missile

SSM Surface-to-surface missile

Missiles During the first World War, the Germans had a gun called Big Bertha, but apart from this, at least in modern times, all guns and torpedoes seem to have been identified only by their maker's name and a mark number.

Almost all missiles, however, have names of their own, ranging from the faintly apocalyptic Israeli 'Gabriel' to the rather unfortunately named British 'Sea Slug'.

Those referred to in this book are listed below:

CROTALE French naval SAM; solid fuel rocket; 10 mile range

EXOCET French naval SSM; solid fuel rocket, sea skimmer; 26 mile range

HARPOON American long-range SSM; solid fuel rocket with turbo-jet cruise; 50 mile range

IKARA Australian Anti-submarine missile/torpedo; 13 mile range

OTOMAT Franco-Italian long-range turbo-jet propelled SSM; 112 mile range

SEA CAT British lightweight point-defence SAM; rocket-propelled; 3 mile range

SEA DART British area-defence SAM/SSM; rocket booster with ramjet cruise; 25 mile

SEA KILLER Italian sea-skimmer SSM; rocket propelled; up to 13 mile range

SEA SLUG British area-defence SAM; rocket propelled; range @ 25 miles; superseded by SEA DART

SEA SPARROW American SAM produced by Raytheon and other firms within NATO

SEA WOLF New British lightweight point-defence SAM, replacing SEA CAT

GENERAL INDEX

INDEX OF SHIP NAMES, TYPES, AND CLASSES

GAS TURBINE INDEX

Engines are listed under their more usual designation, which may be an engine name, letter(s) and number, or maker's name.
The maker's name quoted is the maker at the time referred to in the text. Where the engine in question has since been manufactured by
another company, following a merger, the new company's name is shown in brackets.

A Specification of an Engine for using Inflammable Air for the purpose of procur-ing Motion and facilitating Metallurgical Operation ... Metals, turning of Mills for spinning and Engines for turning up Coals, Minerals from Mines of all sorts, stamping of Ores. raising ...

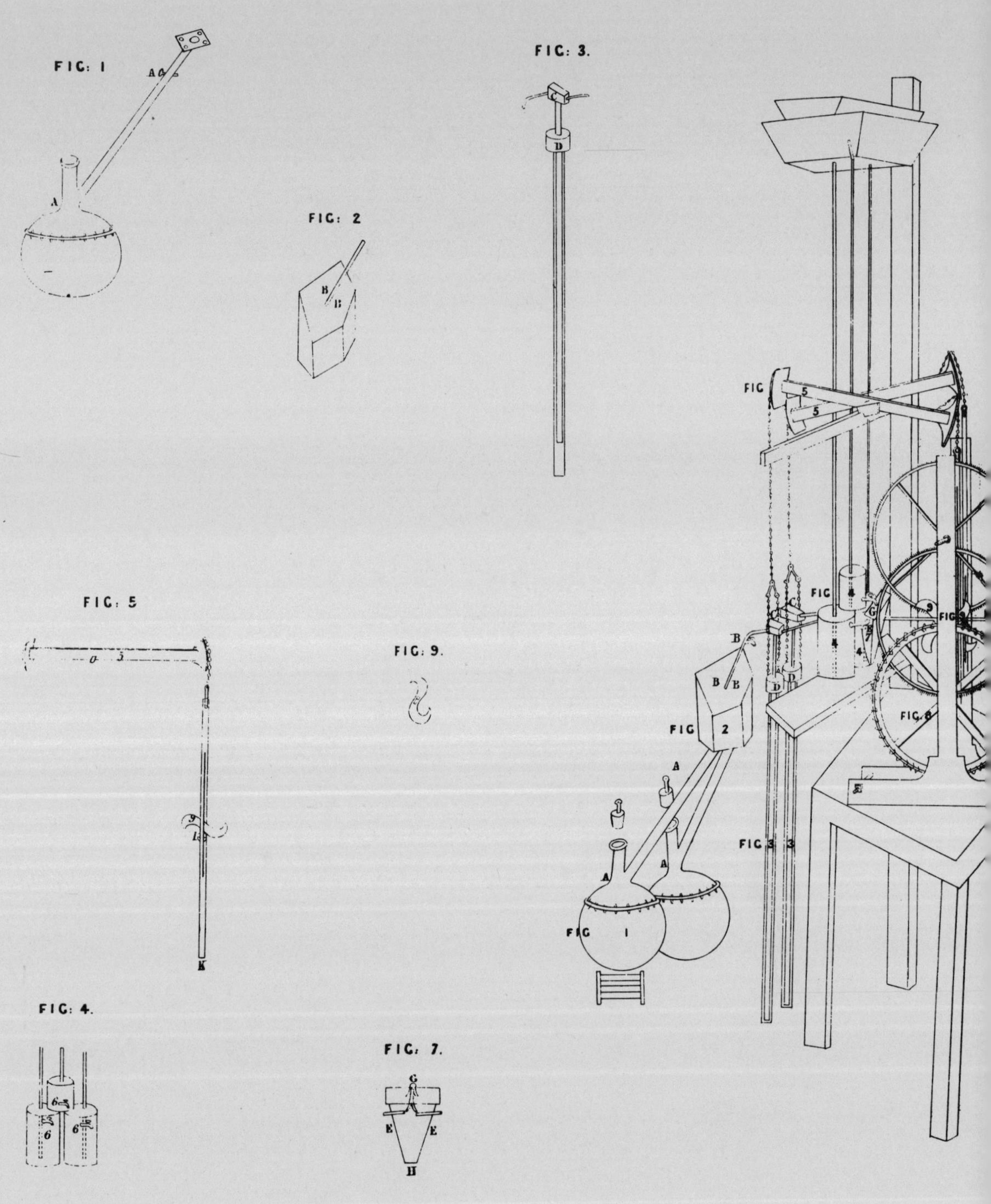

The enrolled drawing is colored

LONDON: Printed by George ...
Printers to the Qu...